RESILIENCE

Survival at the Nexus of Deep-Tech Revolution and Radical Global Change

Dieter Brockmeyer

Diplomatic World Institute, book series, volume III

DWI Publishing, 2026

Cover Photo:
Created with ChatGPT

It is not about predicting the future, but about being prepared for the future."

Pericles

"The future depends on what we do today."

Mahatma Gandhi

Table of Contents

Letter from the President 7

The World at a Crossroads 9

What is Deep-Tech? 20

    All technologies are interconnected 22

    Deep Tech and Energy Supply 31

Four Scenarios 38

    First scenario: data security & disinformation 41

    Second Scenario: The Future of Work & Creativity 48

    Third Scenario: Digital Gold Standard 57

    Fourth Scenario: Law and Ethics 61

Deep Impact 67

    ···on the economy 68

    ··· on society 79

    ··· on climate and sustainability 88

    ··· on national and international relations 98

What Next? 111

    The role of our mindset 113

    Roadmap out of the trap 121

Resilience – Are we ready? 123

Acknowledgments 138

About the author 141

Letter from the President

It is a great pleasure for me to present to you today the third volume in the publication series of the Diplomatic World Institute, once again authored by our cofounder Dieter Brockmeyer, who addresses the still widely underestimated topic of innovation resilience. The waves of innovation triggered by technological progress are accelerating ever more rapidly, reshaping our society in profound and unpredictable ways. We are currently witnessing this with artificial intelligence, which is already causing concern for many people.

Once quantum computing becomes established, with its unimaginable computational power, everything will accelerate even more dramatically. This will not be long in coming. The author illustrates this very vividly, without succumbing to pessimism. Instead, he presents approaches to fostering both individual and societal resilience for discussion.

This will also be the next step in the work of the DWI, where under the term CAMPUS MUNDI step by step we will consolidate all projects related to innovation. These include, in addition to the book series – for which we have now established our own publishing label and to which we are increasingly inviting other authors – the innovation podcast Today & Tomorrow by Diplomatic World, and the academy recently launched with international partners, as well as the Wholistic World Innovation Trophy – or The TROPHY– which was awarded for the fifth time in 2025 and has undergone a comprehensive relaunch, now featuring its own dedicated website. The next module will be a discussion forum, where we will invite committed thought leaders.

Even now, we are open to suggestions and criticism and look forward to engaging with you.

Sincerely

Barbara Dietrich
Publisher Diplomatic World Magazine
President Diplomatic World Institute

The World at a Crossroads

Artificial intelligence (AI) and other advanced technologies grouped under the term *deep tech* have the potential to mitigate, or even solve, many of the world's most pressing problems, such as the effects of climate change. But since the surprise launch of ChatGPT in November 2022, which has already profoundly transformed our society and continues to permeate more and more areas of our lives, uncertainty has spread and at times even panic.

Some prominent scientists and AI pioneers have gone so far as to warn that this technology could enslave us, or even bring about the end of humanity. All this, they say, could happen within just a few years. At the very least, we are told, most of us will lose our jobs. Such predictions stir deep unease in a time already marked by instability and upheaval.

The World at a Crossroads

What unsettles many people even more is the feeling that there's nothing we can do to stop it which only reinforces a growing sense of helplessness. Yet if we want to harness the almost limitless potential of these technologies for the good, we must learn to face our fears, to cultivate a fundamentally positive attitude without ignoring the risks. That, however, is a monumental task; one that is still gravely underestimated by most.

In this book, we talk about resilience – more precisely, innovation resilience. Some might prefer the term future resilience, which perhaps makes the concept a little more intuitive. Both are closely intertwined with technological progress, our societies, and, though often indirectly, with education. Education fosters understanding and tolerance, and forms the foundation on which our communities are organized. In turn, that organization enables resilience.

"Resilience," often described as mental strength or psychological elasticity, is the ability to overcome difficult life situations, crises, or even traumatic events without lasting negative effects and to recover from them. It is the capacity to adapt to changing external circumstances while remaining functional." That concise and fitting definition was generated by Google's Gemini AI. I chose deliberately and with some curation to quote an artificial intelligence here, and I will continue to do so throughout this book, with clear attribution. After all, learning how to engage thoughtfully with these new tools is itself one of the topics at hand.

When we speak of innovation resilience, we do not mean resistance to innovation – that would be resistance or, in stronger form, defiance. Neither is possible, since innovation is omnipresent and largely beyond individual control. The changes it brings, personal, societal, and economic, are simply realities we must learn to navigate.

We urgently need this specific form of resilience as a response to an ever-accelerating, technology-driven transformation of our global society. Yet this is only the beginning. Innovation resilience is a prerequisite for safeguarding the future of humankind.

We must preserve our planet so that we can continue to live well here. It's actually a simple equation. We now stand at a crossroads. Climate change, man-made, as far as we can tell, continues to advance. It is up to us to keep its consequences within bearable limits.

Artificial intelligence and other deep technologies have long been transforming the way we live. As always, it comes down to us, to what we allow, and to what we choose to do with it.

The World at a Crossroads

Technological progress is advancing at an ever-increasing pace, transforming companies and societies alike, leaving us with hardly a moment to pause and catch our breath. The invention of the printing press in the mid-15th century and the steam engine in the 18th century were comparable turning points. Yet while it took roughly two centuries for the first of these innovations to spread widely, the second achieved the same transformation in only a few decades.

Even then, the waves of disruption were severe. However, they came with long intervals in between, and their effects were felt only locally. The Silesian weavers, for instance, initially suffered greatly from the arrival of mechanical looms. Yet that hardship remained confined to their region and gradually resolved itself: some left to seek work elsewhere, others adapted to new trades.

Today, everything happens simultaneously and everywhere. What's more, before we've even adjusted to one wave of change, the next one is already breaking.

Take a recent example: newspaper publishers around the world are struggling to find new business models that can sustain professional journalism. Advertising revenue continues to decline as audiences shift to social-media channels, search engines, chiefly Google, and the growing number of independent bloggers online. The influence of traditional news outlets has dropped dramatically.

There are promising ideas emerging within the publishing industry, but no real solution yet. And already, bloggers and media outlets in the United States are reporting that, at the beginning of 2025, traffic to their websites collapsed, sometimes by as much as 90 percent. The first bankruptcies have already occurred.

The reason: news websites and blogs rely heavily on search queries through Google, which drive traffic to their pages and generate revenue through ad-sharing models. But user behavior has changed radically with the rise of artificial intelligence. ChatGPT and other systems now deliver fully formulated answers to most questions. While they cite their sources as links, most users are satisfied with the AI's summary and rarely click through to the original content.

The result: significantly fewer visits to news sites and an even steeper drop in advertising revenue. Some analysts already predict that by 2028, the entire business model of monetizing web traffic could become irrelevant.

Even if such forecasts should be treated with caution, each shift forces us once again to readjust entirely to a new situation.

The World at a Crossroads

What we increasingly lack is time, time to breathe, to adapt, especially since these changes never remain confined to business models. Their effects ripple through society as a whole.

Long before ChatGPT, we have been confronted for decades with ever-shortening innovation cycles with profound implications for our daily lives. Today, we meet our life partners primarily online, on dating platforms. According to reliable industry estimates, around 60 percent of new relationships in the United States were initiated online in 2024, while traditional ways of meeting people are increasingly marginalized.

In "real life," making connections is becoming more difficult. On public transportation, for instance, it is now almost impossible to strike up a conversation with someone. Most of us are absorbed in our phones, listening to music or scrolling through social media. Even the way we work has changed dramatically over the last generation, and artificial intelligence is accelerating this transformation even further. The next wave is just beginning to build.

Disinformation, often labeled as "fake news" or "deepfakes," poses another threat to our society. Which sources can we trust today? How can we detect deepfakes and maintain confidence in what we see and read? These are just a few examples.

The World at a Crossroads

Elon Musk, the now-controversial billionaire and tech visionary, has suggested that AI could eventually make all of us unemployed. Bill Gates, Microsoft founder and philanthropist, believes that only three occupational groups may remain in the future. Meanwhile, the next generation of supercomputers, quantum computers, is gaining traction. It is only a matter of a few years before their immense power becomes widely available. Tasks that would take the current generation of supercomputers more than 40 years to solve could be completed in minutes, or even seconds, by this new generation.

This will create enormous security challenges. The development of AI itself is also accelerating exponentially. Encryption methods considered completely secure today may soon be obsolete. If we fail to respond quickly, even top-secret dossiers could be exposed online tomorrow. Solutions are being pursued, quantum computing offers some promise. But the question remains: will it be fast enough? Given the leverage that new technologies provide, even relatively small vulnerabilities could cause enormous, globally felt damage.

At the same time, progress is accelerating into the unimaginable. Problems that once seemed centuries away from being solved might now be addressed within a surprisingly short timeframe. Humanity can venture into realms that were previously unthinkable.

The scenarios this opens up are as fascinating as they are unsettling. I recently had a private conversation with someone regarded as a visionary in this field. He is absolutely convinced that AI can finally achieve the humanism that has so far failed. His model envisions a government controlled by AI, based on Singapore's analogue approach. According to him, AI is completely incorruptible and logical. We would, therefore, submit voluntarily to a dictatorship of logic.

This idea, frankly, scares me. Who decides the rules by which this "humanism" is monitored, and who defines what humanism even means? China's Politburo likely has a very different vision than we do in Europe. The differences in perspective could be vast.

Other unresolved questions remain: Can AI itself be monitored? It is self-learning. What happens if it "optimizes" in the wrong direction, or if the person tasked with supervising it abuses their power? And then there is the question of acceptance. Why do many Catalans, for example, seek independence from Spain, even though they are relatively well-off there? Simply put, because they feel patronized. A similar dynamic could emerge under an AI regime: the more unavoidable it seems, the higher the

The World at a Crossroads

likelihood of rebellion. Sound familiar? It strongly resembles popular science-fiction films, from The Matrix to Terminator.

All of this may seem distant. Yet with the pace of change, we could confront these problems far sooner than we anticipate. And concerns about my own job are only the tip of a massive iceberg. Growing uncertainty is already destabilizing society. We can see that this trend will intensify. Coupled with the challenges posed by climate change, which experts warn could trigger massive displacement, the outcome could swell into a monumental tsunami.

This is a worst-case scenario – the gravest possible outcome, which fortunately does not have to occur. But we must remain vigilant. Unstable societies create ripple effects internationally.

At the Diplomatic World Institute, we have been exploring how to address these challenges since our founding in the summer of 2019. We launched the Wholistic World Innovation Trophy which was awarded with its new subtitle The TROPHY for the fifth time in November 2025. Our goal is twofold: to support promising ideas still unheard globally and to stimulate discussion around the concept of innovation resilience.

This is also the aim of our books, podcasts, and the academy, which has just launched this course in collaboration with our partners. Soon, we will consolidate all of these initiatives under the name CAMPUS MUNDI on a single platform. The focus is not on providing ready-made solutions, but on fostering international exchange about potential approaches. No single person or institution can tackle these challenges alone.

It is about equipping ourselves to be future- or innovation-resilient. We are convinced that this is a core task for the coming decades. It also involves accepting the responsibilities that come with these changes. This is also the central theme of this book.

Achieving this requires several conditions: a positive attitude toward innovation and an open approach to change. Neither of these comes automatically. Society will continue to evolve, at an ever-faster pace. Education is the only way to accommodate this. Humans are not naturally equipped to look too far ahead, and our ambition to gain an advantage over others carries both benefits and risks. Ambition can quickly turn into greed. This must be counterbalanced at the societal level through awareness-building and appropriate educational foundations. Humans need a comfort zone. When that is lost, we feel uneasy. In the worst case, this will lead to the consequences described earlier.

The World at a Crossroads

In the following chapters, I will attempt to outline ways to prepare people, to bring them along positively, while preserving this crucial comfort zone. This takes time, and many believe we no longer have it. Yet history has repeatedly witnessed highly pessimistic phases that, in retrospect, proved to be exaggerated. We can hope the same will hold true this time.

Our existential problems are global. Solutions can only be found collectively, across borders. New technologies may complicate this, but they can also form part of the solution. To anticipate one argument: only a market-based approach will work. This is the most efficient way to implement solutions quickly and on a broad scale. I will explain why I am convinced of this throughout the book.

The world's problems and the technologies involved have become far too complex and cross-border to be addressed by any single entity. The world has already become deeply interconnected, a direct result of technological progress. We communicate worldwide in real time, and from Frankfurt where I live, I can reach nearly any point on the globe within 24 hours. When everything is so closely linked, problems can no longer be viewed in isolation.

Today, we are laying the foundations for the world of tomorrow. The legendary Indian civil rights leader Mahatma Gandhi was right when he observed: "The future depends on what we do today."

We can only predict the future to a limited extent, but we can – and must – prepare for it. As Pericles, the statesman and philosopher of ancient Athens, noted around 500 B.C.: "It is not about foreseeing the future, but about being prepared for it." Both insights are prerequisites for fulfilling our responsibility to the planet and for building innovation resilience.

What is Deep-Tech?

First, we need to define what this term actually means. Deep-Tech, short for Deep Technology, refers to technology-based

What is Deep-Tech?

innovations that are founded on scientific breakthroughs or complex engineering achievements. These technologies are characterized by deep research, long development cycles, and significant impacts on society and industry. Typical Deep-Tech areas include artificial intelligence, quantum computing, biotechnology, robotics, new materials, and advanced semiconductor technology, explains ChatGPT.

I recently spoke with an German investor who strongly advocated for a much narrower definition. He wanted to classify only very specific categories, such as profound genetic modifications in medicine to cure rare diseases. Everyday technologies, he argued, should not be included – even though they are the base for the fundamental principles for the breakthroughs in the spectacular fields favored by the investor. Currently, the much-discussed artificial intelligence is just one of the relevant technologies, albeit a highly impactful one. Here, I focus on the core technologies. Biotechnology, new materials, and semiconductors are only touched upon briefly, as they are relevant only in specific, albeit important, areas.

What is Deep-Tech?

All technologies are interconnected

It is important to distinguish: artificial intelligence (AI) works in the background and is not perceived as such by users. Many people, for instance, have long used online translators. They seamlessly integrated into individual workflows and were regarded as conveniences rather than AI. Only when ChatGPT made a well-timed public appearance did many realize the scale of the potentially disruptive changes.

Already today, according to the industry platform Technology IG, students at Harvard and MIT are dropping out – driven by both growing fears and new opportunities surrounding AI. Experts at OpenAI and Google DeepMind predict that AGI, or Artificial General Intelligence, could become a reality before 2030. This is an AI that can think, learn, understand, and solve complex problems across different domains in a flexible and autonomous manner – beyond or similar to human capabilities.

While this academic exodus is not yet a broad trend, some students leave to work on AI safety, contribute to regulation, or launch startups. "I was afraid I wouldn't live to see my degree – because of AGI," Technology IG quotes a former MIT student now working in AI safety. "If we continue like this, AGI will lead to the extinction of humanity."

What is Deep-Tech?

Many, however, are not driven by panic but by economic goals. "According to a recent survey, half of Harvard students believe their chosen professions could become obsolete within a few years. 'If your field is automated before the end of the decade, every year at university is one year less in your career,' says a Harvard AI safety expert. Others see a once-in-a-lifetime opportunity. Some students who dropped out have founded startups now valued at billions of USD. 'There is only a limited window to get behind the wheel,' says another dropout and founder."

AI is already reshaping our lives more fundamentally than any previous technology. And we are still at the very beginning of this development. The pace is most evident in robotics. In China, humanoid robots are already entering mass production and are being considered as universal household helpers. But that's not all: an automobile factory is being built where humanoid robots will work. German manufacturer Mercedes has also announced plans to deploy such robots in its Berlin plant, placing the company at the forefront of development. Thailand has gone a step further, unveiling a humanoid police robot designed to detect and neutralize threats at large events. The list of examples grows every month.

What is Deep-Tech?

Elon Musk expects his robots to be used widely in households within about 10 years, priced around $10,000 each. By 2026, one such robot is slated to travel aboard a rocket to Mars, supporting Musk's plans for a crewed mission roughly 10–15 years later. The development has already gained immense momentum. Robotics is not only advancing rapidly; for many, it also seems threatening. In Japan, scientists have created artificial skin from human cells, allowing robots to smile and regenerate injuries like humans.

Robots are AI-driven. Two years ago, such rapid development was unimaginable. Boston Dynamics, owned by Google, already showcased remarkable bipedal robots most people have seen the viral YouTube videos. These robots could move normally and were even more stable and resilient than humans, but their abilities were limited and they were bulky, with short battery life. Recent videos show that these robots have become far more refined, human-like, and apparently more enduring.

Robotics is not just humanoid. Drones and autonomous vehicles are AI-driven and can be seen as specialized robots. We already see transformative impacts in various areas. In China, a 180-meter-long dam was built entirely without human hands, using 3D printing and AI-controlled robots. This method is expected to expand into road and airport construction.

What is Deep-Tech?

AI penetrates all areas of life. Doctors can already diagnose certain cancers faster and more reliably than ever before. In other diagnostic fields, massive progress is also evident. MIT in Boston developed an AI that deciphered a 4,000-year-old Mesopotamian language in one hour. It solved a problem that had stumped experts for decades. These are just a few examples, and countless others exist.

However, AI can also enable pervasive surveillance. Facial recognition in cameras is so advanced that traditional masking techniques are easily bypassed. While this can significantly enhance public safety, it can also be used to suppress undesirable opinions.

The societal impact is already evident. Anyone can launch campaigns of revenge or manipulation against individuals, companies, or social groups with minimal effort, potentially even causing international conflicts. Creating such campaigns is becoming easier and is disrupting traditional media, as AI can generate high-quality film sequences in cinema quality without crews or actors. AI-generated images, videos, and audio – so-called "deepfakes" – can be virtually indistinguishable from reality. They can reproduce voices and lip movements, placing individuals in any context. Such fakes, easily shared via social media, can cause significant damage before detection, for example,

What is Deep-Tech?

disrupting peace negotiations. Strategies to handle deepfakes remain largely inadequate, and international solutions are urgently needed, especially as creation tools become cheaper and more advanced.

Broad educational approaches are urgently needed to enable people to recognize and contextualize disinformation. Social media and algorithm-driven content are early examples of such influence. In the future, even small, targeted groups can be precisely reached. Both true and fake news can be strategically deployed, accelerating disinformation. The full impact on our lives is only beginning to be understood.

AI agents can already automate virtually all areas of life. Microsoft predicted that by the end of 2024, up to 30% of software programs would have been written by AI. Task delegation to AI is becoming increasingly simple. Previously, prompt engineering required expert knowledge to yield good results. However, the next-generation Gemini AI platform presented at Google I/O 2025 indicates a more intuitive direction. Many new features were hailed as revolutionary, and command input became more user-friendly: "The future of AI is clearly collaborative and intuitive, and our latest announcements reflect exactly that," posted VP Engineering and Regional CTO Cloud Security Wieland Holfelder on LinkedIn.

What is Deep-Tech?

Studies from early 2025 indicate that generative AI is already used by young users as a "second brain," even influencing fundamental life decisions. Other studies warn of potential quality loss: AI increasingly encounters content already written by AI, creating a self-referential loop that could hinder genuine qualitative progress. As monetization via clicks and search engines becomes harder, generative AI consumes the very resources it relies on. Solutions are urgently needed.

Despite AI's massive impact, many in society have not fully grasped its implications. AI simplifies life but displaces jobs, creating uncertainty. Adaptation periods are shortening. Initially, opportunities and risks are balanced: every new technology can serve the common good or selfish, even criminal, purposes. The latter group often acts faster due to fewer scruples.

AI's true power emerges in combination with other high technologies, like blockchain and quantum computing. Blockchain is disrupting global finance and is poised to become a universal transactional standard, providing full traceability. Quantum computers – again, as mentioned earlier – will accelerate research and challenge cybersecurity. Other areas, such as virtual and augmented reality and the metaverse, will also benefit. The blockchain will play a pivotal role both online and offline. Bitcoin is now firmly established as an investment. Central banks could lose relevance in the global monetary system, at least according to some pioneers.

The technology is ripe for surprises, as the rapid value surge of novelty coins like TrumpCoin and MelaniaCoin demonstrated, driven solely by popularity, not intrinsic value. In contrast, Bitcoin is programmed for stability, which it has largely maintained despite volatility. Trump has announced using Bitcoin in the U.S. currency reserve, starting with confiscated crypto assets. Legislative initiatives in the U.S. have further entrenched cryptocurrencies, but effects on Bitcoin remained moderate with a sharp setback started late November of 2025.

Currently, stablecoins dominate financial discourse. The U.S. administration has introduced legislation enabling instant, low-cost global transactions directly between wallets, bypassing intermediaries, but making movements fully transparent. Critics warn this opens the door to surveillance. As Graham Cooke, CEO of Brava Labs, notes, the blockchain aids law enforcement: "Investigations that would have taken years via traditional banks were resolved in hours thanks to the public ledger." Traditional banking cannot compete with this speed and transparency. Regulatory frameworks now require stablecoin issuers to hold 100% reserves, undergo monthly independent audits, and meet KYC/AML standards.

What is Deep-Tech?

The advantages outweigh concerns about control: "Traditional transfers take 3–5 days, cost up to 5%, and pass through multiple intermediaries. Blockchain is instant, cheap, and fully transparent." Stablecoins also enable real-world applications in the metaverse, smart contracts, and supply chain tracking.

Quantum computing accelerates all this. While currently lab-bound, recent breakthroughs demonstrate enormous potential. A quantum computer can solve tasks in minutes that traditional machines would take billions of years to complete. The implications for online transactions are obvious: without rapid adaptation of security, all online activity could become vulnerable. At the same time, research and application development are drastically accelerated, enabling faster solutions to global challenges.

The societal impact will be felt faster than ever. Whereas miniaturization of classical computers took decades, quantum computing will compress this timeline. Many applications will be cloud-based initially, making quantum technology seamless in everyday use. Some quantum computers now even fit on a desk at room temperature, though still expensive.

Medical progress is already immense, much driven by these technologies, as are implications for chemistry and biology. Quantum technology, first explored in quantum computing,

What is Deep-Tech?

could reshape our understanding of physics and reality itself, raising existential and spiritual questions. Concepts like teleportation and time travel have been experimentally demonstrated at the subatomic level. It is not clear yet if this will lead to applications that we only knew for science-fiction franchises like Star Teck. It seems that physics allows it, at least within very complex experiments. The road exploring this further is long and steep and could have a dead end. We don't know yet. However, this is nothing new, for fundamental research opening the doors to new frontiers. The implications are staggering.

I mention all this to illustrate the staggering depth of societal change ahead. Quantum technology has opened Pandora's box, venturing into uncharted territory, as reflected in soaring global patent filings. Societal challenges include the resurgence of conspiracy theories, now merged with elements of esotericism, science fiction, and even quantum science. Deepfakes further amplify these risks.

We are on a wild ride, one we did not choose. This is no ordinary amusement park ride: it offers unprecedented, continuously updated experiences. Seat belts alone are not enough; protective measures must be developed against strong resistance. Yet there is no alternative. We must embrace the ride.

What is Deep-Tech?

Deep Tech and Energy Supply

Let us turn again to less speculative fields. All these developments naturally continue to drive energy demand to exorbitant levels. A foretaste of what happens when electricity actually fails was experienced by the Iberian Peninsula in early summer 2025. There, power went completely out for several hours. The restart took additional hours. Even though the cause was likely not a terrorist attack, but merely a frequency fluctuation triggered by a peak overload, the incident nevertheless shows how vulnerable modern society has already become. There are, however, even more risks, and quite a few experts fear that such events could occur more frequently in the future.

This raises further technical questions, because all known energy sources have their limits. Here, deep tech merges with classical engineering, new tech with old tech. Fossil fuels such as coal, oil, or gas burden our environment and the climate. Renewable energy sources can meet some of these challenges, but whether they are truly sufficient to cover the ever-increasing demand is doubtful. Another problem is ensuring availability during peak consumption: the sun does not shine at night, and the wind is often unreliable in most areas. This means that significant storage and logistical efforts are required to ensure uninterrupted supply.

What is Deep-Tech?

Hydrogen as an energy carrier is considered a solution by many. Artificially produced hydrogen is technically very demanding, both in terms of how it is generated and how it is used. The same is true for the concept of using it as a storage medium for renewable energy. In this approach, surplus wind or solar power is used to produce hydrogen, which is then converted back into climate-neutral energy at night or during periods without wind.

Much more promising is the use of naturally occurring hydrogen. It was long overlooked, yet calculations now suggest that, in theory, its reserves could cover global energy demand for many decades. However, the gas is highly volatile, meaning the quantities actually available may turn out to be significantly lower. Research on this is now underway in many countries and requires substantial effort. Even in the best-case scenario, it will take several years before any deposits can be exploited. Current estimates suggest that this option will not be viable before 2040 at the earliest, which means it cannot simply replace natural gas as a fuel, for example, in heating systems or steel production.

Sticking with the example of electromobility: battery technology in automotive engineering is advancing dramatically and seems far from exhausted. In China, a vehicle battery was presented that charges in 5 minutes for a range of 500 km. This battery also does without the environmentally harmful lithium. Other "traditional" lithium-based vehicle batteries charge more slowly but now achieve ranges of over 1,500 kilometers on a single charge and are said to have a lifespan of 1 million kilometers.

Solar technology appears far more promising in the long term. The trend is toward films and nano-coatings, which can turn roofs, windows, and glass facades into more or less invisible power plants. Japan introduced newly developed solar cells that are said to approach the energy yield of nuclear power. Ingenious engineers are constantly finding new ways to generate electricity. Human creativity in this field is far from exhausted. In Japan, it was even possible for the first time to transmit energy from orbit to the ground using a laser beam. And again, China is not only developing this approach further but pushing it to new limits. There, a structure one kilometer in size at 33,000 km altitude is planned to deliver solar energy to Earth on an unprecedented scale. The annual energy output is expected to match the amount currently extracted from all global oil reserves.

Increasing energy efficiency, reducing consumption in applications, is one way to maintain control over demand. Exciting developments are emerging to reduce the energy requirements of data centers by adopting entirely new approaches to cooling.

What is Deep-Tech?

Even though conventional technology has already made progress, cooling fundamentally requires the same energy expenditure as the computing power itself. Passive cooling can save enormous amounts of energy.

In China, a new silicon-free chip generation has been introduced, delivering 40% more performance than current Intel chips while simultaneously reducing energy consumption by 10%. Considering the dynamic growth in demand, however, this feels more like a drop in the ocean.

A more promising development comes again from China: a nuclear battery the size of a coin has entered mass production, capable of delivering electricity for 50 years. The radioactive radiation emitted by the battery is converted into electricity. Its nominal voltage is currently only 3 volts, making it usable in a limited context, but it is expected to increase soon enough to power devices such as mobile phones.

Progress is also being made in the UK. A carbon-14 diamond battery has been developed there, which could theoretically supply energy for up to 2,500 years, albeit at very low output. Nevertheless, it is interesting for applications requiring a long-term, low-level voltage, such as medical technology (e.g., pacemakers) or space technology. Widespread use at commercially viable voltage and cost is still a long way off, as safety becomes

What is Deep-Tech?

a significant concern at higher voltages. What is still missing, then, is a clean, sustainable energy source that can grow in step with increasing demand.

Fusion energy appears on the horizon as a potential solution. It would be clean, climate-neutral, and virtually unlimited. In recent years, after decades of stagnation, remarkable breakthroughs have been achieved. Whether the now-projected ten-year horizon for the first operational plants is realistic, however, is highly questionable. The technology remains extremely complex, and setbacks cannot be ruled out. Since I first became aware of the concept as a teenager in the 1970s, breakthroughs have repeatedly been promised in "the next decade." It would not surprise me if the same were claimed again. However, positive news is accumulating, and perhaps this time we may actually see a real breakthrough.

Until all approaches generate enough energy to meet continuously growing demand, classical energy sources must hold out. This is why there has been a renaissance of "traditional" nuclear power in recent years, despite the unresolved problem of nuclear waste. More and more countries are building conventional reactors again. However, there is also movement toward small energy units that can be transported in containers and relatively easily installed or replaced. Once only a concept, implementation is now progressing. Thorium, not uranium, is being split, with the advantage of producing less waste that is less radioactive and decays much faster. Even a worst-case accident, like Chernobyl or Fukushima, is no longer feared. Its half-life, the time after which only half the radiation remains, is significantly shorter. After a few hundred years, it is no longer radioactive, and the waste is not weapons-grade. These are all advantages. Bill Gates is also heavily supporting this concept, and China has now brought its first reactor of this type online. Interestingly, China reportedly developed its technology rapidly based on now-released documents from the US, where the topic had long been ignored.

Nevertheless, the waste generated by conventional nuclear power plants remains a problem, as it remains hazardous for thousands of years. The new generation produces less and less toxic waste. The "transmutation" of nuclear waste discussed by scientists to neutralize radioactivity is still purely theoretical. AI and quantum computing could potentially accelerate a proof of concept, but this remains a distant hope. Another interesting development involves a fungus discovered in the ruins of the Chernobyl plant. It feeds on radioactive radiation. The hope is that it could lead to a tool to neutralize nuclear contamination. However, nuclear waste itself becomes a resource, even for everyday applications. The radioactive material, carbon-14, for

What is Deep-Tech?

the aforementioned nuclear batteries is sourced there. All this is promising, but we need to keep in mind that all this is in the very beginning stage, with no guarantee for customized applications.

Overall, the outlook is not entirely negative, provided these positive approaches are not misdirected and reach market maturity quickly and without major delays. Many initiatives may fail for various reasons, this is the normal dynamic that makes forecasting so difficult. The key question remains: will we be fast enough?

Whether deep tech or energy technology, all these innovative approaches intersect and trigger massive changes across society, which can also create uncertainty. To manage this, innovative approaches are again required, demanding a holistic perspective. Our horizon, however, is limited. Only if we acknowledge this can we develop the openness necessary for genuine compromise. We must view new technologies positively, as potential problem-solving tools, without losing sight of the associated risks.

Four Scenarios

There are currently more than enough scenarios, both good and bad, in the face of the shock OpenAI unleashed only a few years ago with the release of ChatGPT. Have we, with artificial intelligence, ushered in the end of humanity itself? Elon Musk has spread that idea. But he is far from the only one raising a warning voice. The CEO of OpenAI, Sam Altman, and a group of no fewer than 376 experts have also warned urgently about the almost apocalyptic dangers and therefore demanded comprehensive, rapid regulation of AI. Even today some reports from the AI front are disturbing. For example, one AI reportedly developed defense strategies on its own when it suspected it was about to be shut down. It began to lie to prevent that. That strongly recalls popular themes from science-fiction literature.

Four Scenarios

So it's not surprising that other critics go even further: "AI will have a destructive impact on society worldwide," a computer scientist from Oklahoma State University declared in an interview with the British tabloid The Sun. He predicts the world population will shrink to only 100 million people by the year 2300 (from over eight billion today). Apart from the fact that such long-term forecasts are completely unserious, his prophecy is overall pretty much nonsense. Such horror scenarios always appear when new technologies emerge, usually spread by attention-seeking provincial know-it-alls. Occasionally, however, there are also voices worth taking seriously, which, given a tool as powerful as AI, is unsurprising. Such negative developments would presuppose that humanity did absolutely everything wrong and allowed technological rampant growth to run free. But that is not to be expected. Caution is of course still warranted.

But there are also supposedly positive scenarios: the former Google engineer and almost legendary futurist Ray Kurzweil promises us eternal life as early as 2030. Nanobots, microscopic robots in our bodies, for example, are supposed to repair cells or stop the aging process. His legendary reputation comes not least from the claim that more than 86 percent of his predictions have been correct. He predicted, for instance, that computers would beat chess champions, which happened in 1997 when the

Four Scenarios

supercomputer Deep Blue defeated Garry Kasparov for the first time. Kurzweil predicts AI will surpass human intelligence by 2029, and by 2045 we could fuse with AI into ultra-intelligent hybrids.

Are these really positive scenarios? Take immortality alone: current advances in medicine are rapid. Here are just a few examples: cancer cells can be targeted directly in the body and either destroyed or cured. Scientists have decoded the secret of regrowing organs and limbs in amphibians and hope to one day apply this to humans. In Japan, clinical trials are already enabling adults to regrow lost teeth. These are not yet widely applicable therapies, but such treatments could become available within a foreseeable time frame.

These developments will undoubtedly continue to raise our statistical life expectancy and potentially extend our capacity and performance, perhaps significantly. Whether this will actually lead to immortality is another question. But if it were to happen, what would it mean for issues such as overpopulation, and how would we respond to life fatigue that might afflict many people over the course of countless additional years gained? Another area of concern: who would get access to these therapies? Would class boundaries harden into an "immortal" upper class and a mortal precariat?

Four Scenarios

What all this makes painfully clear is that, even if we already feel overwhelmed by the speed of technological progress, we must repeatedly remind ourselves that we are still very much at the beginning.

Nevertheless, I would like to develop four scenarios here, less speculative ones, that will predictably influence each of us already today. Of course more could be derived. But I will stick to these examples. I will try to "push back" on these scenarios, i.e., to frame them more broadly than they currently appear in public debate.

First scenario: data security & disinformation

We have already spoken several times about the emerging problems for data security posed by quantum technology in an increasingly digital world already under pressure from cybercriminals. And fake news and deepfakes rightly occupy a large space in the current discussion and are a danger to democracy in multiple ways. Our handling of both, however, remains rather helpless so far. The term "disinformation" is said to be of Soviet origin and has only been in use in Western countries since about the early 1970s. Before that there were a variety of other terms: "rumor," for example, or "canard" when something completely false appeared in the newspaper. Seen in this light, the phenomenon of fake news is by no means new. What has changed are the means by which it can be created and spread in real time across borders.

Two things have unquestionably changed because of social media. While "armchair politics" previously could not spread because they lacked a broad resonance space, the purveyors of such chatter, who felt strongly even back then but could not objectively verify their reach, can now do so. The new communication platforms have thoroughly changed that. The spreaders recognize their power and now tend to overestimate it, because they receive reinforcement only within an "echo chamber" amplified by the platform operators. As a result, their self-confidence grows and they become even louder. On the other hand, we can now clearly see that a strong parallel world exists. This parallel world is often overvalued because of its loud and provocative presence. Yet it is not new, and in fact we therefore have the opportunity, through this visibility, to tackle the problem effectively. Unfortunately, the reaction of those affected has been rather helpless, and measures taken, driven by initial reflexes, again only address symptoms instead of tackling root causes.

Four Scenarios

Jeanette Hofmann, director of the Alexander von Humboldt Institute for Internet and Society (HIIG) in Berlin, formulated a somewhat more differentiated approach at the Frankfurt "forum medienzukunft 2025" of the Medienanstalt Hessen, a local media regulator. She argues that the spread of disinformation is an expression of an erosion of trust in democracy. There is no scientifically robust evidence that disinformation changes behavior or even makes people stupid. While the categories "true" and "false" are of limited use in political discourse, narratives are used "in the gray area of facts, rumors, opinions, or lies" to communicate political action. Evidence, facts, and systemic trust are increasingly being replaced by a strategy of "personalization of trust." Jeanette Hofmann (...) described a "rollback of standards, driven by populism." As a result, trust in democracy falls, and many develop a "great longing for charismatic leaders." The problem, she concludes, is not social media per se, but the mistrust of democracy that finds its way into public view. If not via Facebook and TikTok, she suggested, it will find its way via tabloids and newspapers. It is striking how similar, although with different signs, certain media treatments were in the run-up to the Nazi seizure of power in Germany in the 1930s.

It is also notable that fake news from one camp are taken up by the other simply because they reinforce a discomfort or mistrust of one's own camp. For me it was a clear signal about the

Four Scenarios

possible outcome of the 2024 U.S. presidential election when, just one month before the vote, at the edge of an event I met a smart young entrepreneur from the U.S., a traditionally liberal New Yorker. He told me he had always voted Democratic his whole political life. This time he said he would not go to the polls for the first time because the Democratic candidates were "all corrupt." As proof he cited that victims of a recent devastating hurricane in the southern U.S. were supposed to be given only $500 for the loss of their houses.

Of course that was nonsense! The amount referred to was emergency aid intended to enable the victims to obtain the most necessary clothing and toiletries at first. The entrepreneur had fallen for the propaganda of radical MAGA supporters, even though the accusation was absurd and the fact would have been easy to verify. He didn't care, because it supported his sentiment. The MAGA (Make America Great Again) camp had thus succeeded in infiltrating parts of the opposing political camp with fake news and at least keeping some people from voting. We therefore need completely different approaches to deal with the phenomenon of fake news and disinformation. What is needed is engagement with the issues themselves, not a platform-based response.

Four Scenarios

Something else looks different: portals on the web that are deliberately concerned with spreading outright false news. On the one hand there are so-called clickbait sites whose only aim is reach in order to generate advertising revenue; they deliberately rely on sensational and emotional content that usually has very little to do with reality. On the other hand there are troll factories that disseminate content on behalf of an interest group to create unrest in a specific group or to destabilize society as a whole. These are pursued by law enforcement and their pages are deleted. This approach is not particularly effective, however. These sites are hosted on servers that are difficult to identify and reach, often in countries outside mutual legal assistance treaties. If a shutdown succeeds, they are immediately reinstalled elsewhere and the game starts again. Here, too, the only effective weapon seems to be educating society at large so that fake news are recognized and ignored. In times of social division and growing mistrust in institutions, that is of course a particularly difficult undertaking.

This situation is made more difficult still by technical progress, here with AI and so-called deepfakes. The best-known variant of deepfakes is face-swap, where faces in an image are simply swapped. For example, the face of a global star is inserted into a hardcore porn scene and sold on porn sites. The tools for this are available to anyone, usually in a basic form for free, on the internet and with a few clicks one can create such content. Much of it is harmless, but much of it is not. Even if a global star would never appear in a professional porn film, such videos are at least defamatory, because some people might take the fake seriously. Even more severe is the potential effect of a targeted campaign to destabilize society. Words could be put into the mouth of a head of state or the CEO of a corporation that they never uttered. Diplomatic turbulence or panic on the stock market could easily result. Again, broad societal awareness is the main remedy, but it faces the same hurdles described above.

Data security must also be mentioned in this context. But it encompasses far more than defending against the spread of fake news, as a look at the statistics shows. Year after year the damage caused by so-called cybercrime is rising. In 2024 the damage worldwide caused by it was estimated by Statista at around 9.2 trillion USD.

This figure encompasses a wide range of different offenses: destructive hacker attacks are included as are phishing i.e., the harvesting of sensitive data, industrial espionage, cyberbullying, or copyright infringements, to name just a few examples. The Bitcoin community puzzled over an inexplicable movement of 8,000 BTC in mid-2025. The analysis found it was very likely triggered by a hack. Fake news are part of this, too, even if they

Four Scenarios

are often only one of the means used to prepare or conceal other crimes online.

This shows how important the aspect of data security is. But it is again a double-edged sword. Everything that can be used for protection can also serve the opposite purpose. Quantum technology already heralds the next wave of challenges. On the one hand, existing infrastructure and everything we have implemented to secure it could become obsolete. On the other hand, it promises a security standard based on an entirely new architecture that, as things stand today, is supposed to be absolutely secure, at least for the time being.

The problem is the transition. The risk here is that parts will not be switched over quickly or cannot be switched over (both are likely). We should prepare ourselves for some unpleasant surprises during the transition. But even once the switch is implemented, we must not rely on this feeling of security. Every technology is progressive, and it will again be only a matter of time until backdoors are discovered even in technology that previously promised us absolute security.

One more topic I will only touch on here to raise awareness: just as we secure our data infrastructures with firewalls, we increasingly need to do the same in public and private spaces. Deep tech delivers ever-better possibilities for monitoring these spaces. We all know the example of China, where public spaces are now surveilled comprehensively with cameras. The images are analyzed with complex software, for example for facial recognition, so that almost anyone present in a public square can be identified. The technology is becoming ever more sophisticated. In a lab at the University of La Sapienza in Rome, Wi-Fi in buildings was for the first time used to monitor the rooms supplied with internet in this way. The software detects disturbances in a room's Wi-Fi field and can deduce how many people are in a room and where they are. The technology, called WhoFi, can even identify people fairly reliably from the distortions, at least if their patterns are known. This promises an effective and above all inexpensive method, for example to protect against burglars or espionage. But it can also be used for arbitrary surveillance or intimidation of employees and for the suppression of society or similar ends. The possibilities for both uses are getting easier and pose ever-new challenges for us.

Second Scenario: The Future of Work & Creativity

Billionaire and philanthropist Bill Gates once again attracted attention. In just ten years, doctors or teachers might barely be needed anymore, because AI could perform most tasks in these

Four Scenarios

fields far better. They won't disappear entirely, but their numbers will shrink to only a fraction of today's levels.

This list of affected professions could certainly be extended to include lawyers or notaries. In fact, hardly anyone can truly be excluded from it. Gates even predicts a two-day workweek for most people. Yet he adapts very dynamically to developments. He now predicts that only three professional groups will survive the AI revolution: programmers, energy experts, and biologists.

It doesn't really matter whether this takes ten, fifteen, or even twenty years, it can vary depending on the profession. In many areas, a certain workforce will be maintained for safety reasons, since, as we know, AI still makes mistakes that are more difficult to detect the better the AI gets. There more is potential to cause delay: In England, after the introduction of diesel and electric locomotives, unions had ensured that stokers, those who shoveled coal into the boilers of steam locomotives, had to remain on board the new diesel engines.

The question inevitably arises: what do we do with our productivity, and how do we secure our income in the not-so-distant future if our labor is no longer in demand? There is now an ever-growing group of people who believe that the solution can only be a "universal basic income," financed through an "AI tax" that captures the efficiency gains AI generates in companies. What seems logical at first glance quickly raises questions. For example, how can a person receiving basic provision use their productivity meaningfully for themselves? Otherwise, dissatisfaction quickly arises, creating fertile ground for unrest.

Proponents of the theory cheer that humans could finally pursue hobbies and activities for which there is no time under today's "grind." This is even supported by an accompanying study of the first larger privately funded basic income trial in Germany: The accusation that a universal basic income would make people lazy is unfounded, according to the study's authors. Whether this holds true for larger sample sizes still needs to be proven. Participation in the trial required an application. Participants had to have heard about it and taken action, meaning the sample may not be truly representative of the general population, and likely included highly motivated individuals.

However, this motivation probably applies only to a limited segment of society, while others want or need to remain employed. A friend and international investor saw a solution in e-sports, online competitions where people can spend days and nights in front of screens, competing for small prizes. I also find this approach problematic. The reasons are obvious to anyone.

I believe the solution will consist of multiple building blocks. Traditional craftsmanship, largely without computerized tools and with the charm of imperfection, will regain significance. Of course, this will be a niche in which only a limited number of workers can earn a living. There could be the problem that AI might perfectly replicate the charm of imperfection, making handcrafted work indistinguishable from machine-made products. Yet for now, imperfection still holds its appeal: we see this in the resurgence of vinyl records, which have become so popular among music listeners that not only the old pressing plants have been restored and new ones are produced. Once again, it is a profitable market, albeit a niche one.

At the same time, entirely new professions will emerge that we cannot yet imagine. The number of "coaches", trainers for every life situation, has exploded in recent years, and using one has become fashionable, even if many still think it's unnecessary. Have you ever heard of a "tree coach"? He takes clients into the forest, explain the trees and the forest ecosystem, and bring them closer to nature. The highlight is an overnight stay in a tent in a tree. Apparently, he is quite successful.

When considering the future of work, we must also consider so-called digital workflows, which are becoming far more widely applicable thanks to AI. Repetitive tasks could already be automated before, usually dull and tedious activities. Now, hardly any area is excluded. AI allows integration into virtually all processes.

Digital and AI-driven workflows simplify many tasks but also destroy jobs, not just unpleasant ones, but across the board. They still have disadvantages: if, for example, packages in an online store's shipping department are labeled incorrectly due to a faulty label printer, causing delays and returns, it may take days to detect and fix the error, generating frustration for call centers and customers. Such incidents are currently not uncommon.

Why mention this? Because it shows that, for the foreseeable future, trained human service staff must be maintained to minimize risks of both approaches. At present, this is often sacrificed prematurely due to cost control, as service efforts reduce margins. Customer frustration is largely ignored, not least because all companies currently use these tools, leaving customers with no alternative. This will change, even as AI and workflows improve. Human interaction and empathy remain indispensable in customer dialogue. It is doubtful whether AI will be able to conduct truly deep customer interactions in the near future.

Even though we are currently in a phase where entrepreneurs are exploring the limits of automation, I am convinced that the

Four Scenarios

human factor will experience a renaissance. This will take time because the ever-fueled hype around technology and cost savings must first be satisfied, with increasingly comprehensive innovations rapidly following. Robots will be optimized for their specific tasks in the foreseeable future. A robot waiter may be perfect in service, perhaps even capable of some small talk. But individualized, guest-specific dialogues, such as recommendations for outings or shopping, are far more complex and remain a long way off. Here, humans still have the advantage.

However, there is a limitation. Even empathetic human service staff will not be everywhere. They will exist in high-end gastronomy and luxury hotels, while ordinary customers will be served by machines. For some time, human service will continue to decline, partly because robot waiters will initially fascinate people. But they will quickly get used to it, and something will soon be missing. Perhaps I am too optimistic. Still, I see substantial, though niche, potential. Again: no matter how good AI is at simulating human behavior, it cannot yet match real emotionality and empathy. This represents one of the opportunities I see for the future of work.

Whether this potential can be expanded depends on the economy, on whether it can be made sustainable beyond just climate issues. The more sustainably we shape society, the economy,

Four Scenarios

and the labor market, the greater the possibility for a more humane economy. At least in gastronomy, care, and service, we should not forgo the human factor, despite tasks taken over by AI and robots. Whether it ultimately prevails is our responsibility.

That was a rather long-term perspective. In the short to medium term, the main problem is different: the speed at which AI is changing the job market, far faster than the labor market, especially vocational training, can react. Those starting a degree or vocational training today often do not realize that by the time they finish, their profession may no longer exist, or will exist in a completely different form.

We already saw a phenomenon almost immediately after ChatGPT's launch. Previously well-booked experts suddenly had no assignments, as AI could perform their work faster, cheaper, and for anyone to access. Interestingly, it first hit photographers and layout artists hardest, industries previously considered resistant to AI. AI itself is not creative, we were told; creativity requires humans.

This is true, but the impact was underestimated. Genuine human empathy still beyond AI is critical to creativity. Despite all the AI developers' fantasies, human creativity remains superior. AI only does what we instruct it to do, albeit incredibly fast and

Four Scenarios

precisely. In the long term, that is not enough. So I am not worried about the future, only about the rocky path ahead. We must prepare for it.

Human creativity is the linchpin of our continued success as a species. It is central to the question of the future of work. But it goes further: if tasks and thinking are largely taken over by AI, what happens to our creative abilities? AI is meant to support us, or will it ultimately be the other way around? There is palpable panic in the creative industry. The entire sector must be rethought, says Wesley Ellul, global Web3 entrepreneur and special envoy for startup funding and investment of the Maltese government, in the podcast Today & Tomorrow by Diplomatic World:

"The creative industry is relatively safe, but a traditional designer who constantly replicates others' work has no future, graphic designers who do nothing new, just copy others. Average video editors who only meet minimum requirements are also not secure, AI can do that just as well, if not better. Those who employ genuine creativity are safe. Ultimately, people who combine creativity with AI are future-proof."

At the World Economic Forum in February 2025 in Davos, a list of the most in-demand future professions was presented. The top four were highly technical jobs, such as AI engineers. But

Four Scenarios

fifth place was already creative thinkers. This trend continued: talent management ranked seventh, considering personality types and networks. Every good lawyer knows many people and governments cannot afford to lose such individuals from society. Likely, the state will take care of them. The World Economic Forum predicted this as early as 2020.

But what will the new economy look like if, as some forecasts predict, a billion people lose their jobs? Money in their pockets, but infinite free time? Some will try to fill it meaningfully, others just want to enjoy it. That is why the aim is to bring together people from Web3 and Web2, so they can shape their future instead of leaving it to governments."

The question that arises for many, including myself: how many of these creative thinkers can there truly be space for? Much creative work has been reproductive. How many can break into new areas? Moreover, the creative approach will likely change in the future. I notice this in myself. My approach is intuitive. I develop an idea here, a thought there, then try to integrate it into the project. This integration is not static; it changes with new ideas.

In the case of a book, half the content is already written, two-thirds almost complete before I could even use AI. AI requires a structured approach: outline, direction, content, all predetermined in prompts. It's a completely different process, with little resemblance to classical creativity. The danger is getting sucked into the digital workflow. Publishers using AI-optimized processes can not only save costs but bring books to market much faster.

Books on current topics will thus be available sooner, while still relevant. But this also pressures the author to adapt to AI's pace. A mainstream effect arises, from which it is hard to distinguish oneself. Plus, there is competition from AI-produced content, rapidly created and independently published on mass platforms. Creative quality struggles to stand out, let alone be noticed, getting lost in the mass. This is a tremendous loss.

Third Scenario: Digital Gold Standard

Marc Friedrich is a German financial expert, somewhat controversial, who presents himself and successfully with the public as a crash prophet. His analyses often strike a chord with the times; his conclusions – subjectively – are not always accurate. Like many in the crypto sphere, he is a passionate advocate of Bitcoin and sympathizes with a new global reserve currency to replace the US dollar, initially a hard currency, either pegged to gold or to Bitcoin.

Four Scenarios

"Many believe that whoever embraces Bitcoin first will gain the new reserve currency, and that the US has recognized this, since it does not want to lose its dominant role. The approval of Bitcoin ETFs could be an indication that this is true," he lectured at an event in Frankfurt. Perhaps. But the question remains: is this really such a good idea?

A few months later, this vision seemed closer when former President Trump declared Bitcoin one of the official reserve currencies of the US. Meanwhile, the US has taken further steps toward general integration of DeFi, decentralized finance systems based on blockchain technology, into the traditional financial system.

There are good reasons to look for alternatives: progressive inflation, which erodes private wealth, or the extremely high national debts, which can only be controlled through inflation. These factors repeatedly trigger severe financial crises worldwide. Thus, predictions of a near-future collapse of the global financial system are increasing. Until the Nixon Shock in 1971, the US dollar was pegged to gold at $35 per ounce. This arrangement, the Bretton Woods system, allowed countries to exchange their dollar reserves for gold at the US central bank, providing relative stability for the dollar and associated currencies. Nixon unilaterally suspended this mechanism to cope with financial pressure from, among other things, the Vietnam War. That was effectively the end of the gold standard, and since then, we have seen increasingly rapid loss of purchasing power in the dollar zone.

Undeniably, Bitcoin has become more valuable despite volatility, especially because the number of coins is strictly limited and mining is increasingly difficult. Bitcoin is therefore a "hard" asset, in some areas even harder than gold, as predicted in Satoshi Nakamoto's manifesto. However, Bitcoin pioneers lament its increasing integration into traditional financial structures, which erodes its independence. The world's largest asset manager, BlackRock, already holds the largest Bitcoin reserves. Central banks are also working on issuing their own digital currencies, independent of Bitcoin, based on national fiat currencies.

A crucial distinction between Bitcoin and gold is often overlooked: farmers occasionally still find gold buried hundreds of years ago, but Bitcoin cannot be "found" in this way. Bitcoins are data and can be deleted for any reason. A farmer might only find the remnants of a USB stick or hard drive, too damaged for recovery. Gold, once released into circulation, remains; data, however, is volatile.

Four Scenarios

We live in increasingly uncertain times, and many fear that we are entering a phase similar to the transition from the Pax Britannica to the Pax Americana, which may now be ending. Evidence supports this theory: a long period of crises, even wars, could follow. Hybrid attacks on critical infrastructures like electricity and communications networks are already feared. AI and quantum computers may establish unprecedented security standards, but they also expose vulnerabilities we cannot yet imagine. It will be a race that could easily trigger larger and prolonged system failures in critical infrastructure.

In the event of even a limited nuclear conflict, electromagnetic pulses (EMIs) could occur, for which we are only partially prepared, potentially frying circuits and returning entire countries or continents to a pre-electric age. Data, including digital wallets, could suddenly become inaccessible or even permanently deleted. While Bitcoin's structure offers protection, lost value is redistributed among remaining assets, if wallets become inaccessible or holdings shrink, this effect is limited in these situations described.

Even if currently considered extremely unlikely, such a scenario would be catastrophic for Bitcoin and digitalization overall. Even a less severe scenario could cause massive loss of trust, hardly survivable for the crypto ecosystem. Some in the scene consider such a scenario plausible. "I'd rather not think about it," said a crypto pioneer. But ignoring it is exactly the wrong approach.

Fourth Scenario: Law and Ethics

A business partner introduced me and my latest book CAMPUS MUNDI, saying it was partly about ethics. At first, I wanted to object. I hadn't used the word ethics once. But then I realized she was right. Everything I wrote is consistently grounded in ethical principles.

Laws attempt to channel society according to ethical principles. Law and ethics are inseparably the foundation of our coexistence and the functioning of societies. Of course, there are exceptions. I do not deny that. Legislation and jurisprudence in Germany's darkest times (not the Middle Ages) were based on ancestry and Aryan status, linking legal rights solely to lineage and loyalty to the regime. This too was founded on definitions of ethics and rule of law, though vastly different from what is commonly understood.

We are not talking about religious or ideological delusions. Let us assume a relatively liberal society. Even there, constitutional controversies are rooted in ethical interpretation. Consider, for example, the debate on abortion, which is controversial in all industrialized societies. I will not take a stance, only describe. The conflict between free self-determination and responsibility for (unborn) life cannot be resolved easily and is deeply ethical. Yet this is far from the only topic that will challenge ethical discussions in the future.

Consider work and the economy. Work should allow a life without major worries, and the economy should serve humanity. In times of precarious employment and shareholder value, focused solely on profit maximization, we have strayed far from this ethical principle. AI, despite grandiose promises, will likely exacerbate this trend, unless we actively counteract it. The list of examples could be extended significantly.

Ethically translating such questions into laws is fundamentally difficult. Technological progress only complicates matters. When does AI develop consciousness, and will it then have human rights? Philosophical hypotheticals are rapidly becoming real. Society is challenged: what is ethically justifiable, and how must it be codified in law? Consider a seemingly banal example discussed with a renowned German intellectual property lawyer:

"We will see systems delivering content individually tailored to us, with no way to trace why we received it," he said, asking: "Why did I get this? Who uploaded it? Is it true or false?" We

Four Scenarios

would be trapped in a bubble fed by personalization, using collected data from countless sources, and AI-generated content added individually. I know this user only cares about sports, they then receive only AI-generated sports content with relevant ads, automatically. The advertiser pays per completed purchase. All happens in the background. We ask why we see this ad, yet cannot track it. Regulators cannot either, they see different content. Dissecting the algorithms that combine and analyze this data will be impossible." Questions about ethical advertising behavior are at risk of being completely ignored, and similar scenarios apply in other domains. These issues are already highly relevant.

These aspects highlight the scope of the challenge, demanding rapid legislative action. The faster, the better. Yet it seems almost impossible. Europe sees activism and asks the right questions, but answers rarely look forward and could worsen problems long-term. A high-level expert commission in Germany once recommended bureaucratic streamlining. Applause? Too early. The main goal was accelerated digitalization to improve administrative efficiency.

At first glance, this sounds good, but it suggests that bureaucracy is being replaced with digitalization. Rules such as e-invoicing, however, remain overly complex. Despite lip service, detailed micro-regulations still exist, especially in digitalization. Shortly after the report, Federal Labor Minister Bärbel Bas in Germany proposed the "Tariff Compliance Law," including a new authority to oversee corporate compliance with agreements on Christmas bonuses and other matters.

The consequence for companies is clear: more bureaucracy. Perhaps I am pessimistic, but more bureaucracy means more opacity, even when increasingly hidden behind AI. Errors, when they occur, are buried within AI processes, barely traceable. This is symptomatic of the overall approach: it spreads like a cancer, paralyzing businesses and citizens alike. The citizen is excluded from the debate, as if the economy could separate from society.

EU regulation, often seen as excessive, stems from Europe's concept of freedom. Unlike the US, which emphasizes individual freedom, Europe prioritizes societal freedom, requiring occasional individual restrictions for the community. This legal definition directly affects legislation. In globalization, this creates conflicts with the US, as approaches are hard or impossible to reconcile. Regulations proliferate to ensure a "level playing field," not in general equality but to standardize rules where comparable.

Four Scenarios

Strict application leads to increasingly dense regulation, seen daily in EU laws, particularly in innovation sector, harming users and businesses. US conflicts are not new, though more visible today due to tone changes. Globally, legal systems struggle with analog thinking, treating symptoms rather than reforming fundamentals. Law should be an enabler, without losing sight of risk prevention.

Currently, legal structures and courts remain rooted in analog thinking. Both systems are highly detailed, repeatedly revisiting matters despite central regulation. Fraud is fraud, whether on the street or online, yet new laws are often created unnecessarily. Internet law, initially seen as lawless, is now painfully slow to update. Reforms are outdated by publication. We are trapped in endless loops; what is needed are concise, forward-looking laws accommodating future applications. Less, clearly focused and open to the future, is more than micro-detailed regulation.

I expressed this spontaneously during a panel, receiving enthusiastic nods from Web3 lawyers. Later, a traditional lawyer criticized it, correctly noting that practice follows precedent. Each ruling reduces interpretative flexibility. US law relies heavily on precedents; elsewhere, less formal but similar. Courts must adopt more openness. Reducing regulatory density need not

Four Scenarios

conflict with a "level playing field", it simply creates flexibility, without changing scope. Balancing this remains challenging, with overregulation creeping subtly. Reacting to every incident and tightening laws reins in citizens gradually. The key challenge is shifting legislators' and judges' mindset.

Rapid action is essential to ensure legal certainty in a constantly changing environment. Accelerating technology raises fundamental ethical questions that must be debated and, where necessary, codified legally. Daily legal minutiae easily obstruct this. The consequences are immediately felt, widening societal divides. This must be prevented.

New technologies offer extraordinary opportunities if used positively, demanding a clear ethical and legal compass currently lacking. Ethical questions multiply. This brings us full circle: as life expectancy increases dramatically, questions of life's end arise. If my body continuously renews, when may I decide it is enough? Who may intervene? Will the state claim this decision for its citizens?

This is not entirely new. Some countries allow or tolerate self-determined death, e.g., terminal illness, while others struggle or reject it outright. The topic is complex; clear answers are rare. Technological and societal change raises fundamental, controversial questions capable of destabilizing the world unless viable solutions are found early. We must seize this opportunity.

Deep Impact

The scenarios I just outlined have taken a close look at individual segments. It is evident that their effects extend far beyond themselves. They lead to lasting changes across all areas of society. That is why I will now attempt to broaden the perspective – though again, this can only be illustrative – looking at the economy, society as a whole, climate change as a central challenge of the coming decades, and finally at politics and international relations, which must manage and shape all of this.

So again, we are dealing with four segments, which are closely interconnected. When I talk about bureaucracy, for instance, it slows the pace of innovation and the economy overall, while simultaneously constraining society, slowing its adaptation to changing conditions. It also delays and increases the cost of transitioning to sustainable technologies and thereby hampers climate protection. Furthermore, it prevents rapid responses to shifts in the international balance of power. Bureaucracy is only one example of factors influencing these areas, but they also interact with each other, as we will hopefully see in the following pages.

Let us begin our analysis with the economy, not least because it is usually here that technical innovations are implemented, which then have broad effects on everything else. Innovations change processes and become products that are distributed widely through the economy and thus exert transformative influence.

···on the economy

The economy responds particularly directly to changes, at least when they are beneficial to existing business models. Anything that seems to promise short-term efficiency gains or cost reductions quickly finds its way into everyday corporate practice. More complex innovations, which challenge and change existing business models, face greater resistance, because there is

Deep Impact

usually no "proof of concept", no guarantee that it will actually generate revenue. This slows adoption and, in critical situations, can create the impression that the economy is incapable of delivering. Yet there is always a tipping point, after which new models are broadly accepted and gain tremendous momentum. The rapid post-World War II resurgence of Germany, the so-called "Wirtschaftswunder" (economic miracle), was only possible within a capitalist system.

It is already becoming clear that the same will happen on a global scale regarding climate change. China continues to invest massively in coal-fired power plants to satisfy its ever-growing energy appetite. At the same time, renewable and alternative energies are gaining ground and have grown faster than fossil fuels in recent years. Projections for 2024 suggest that CO_2 emissions have declined for the first time, despite the continued rise in energy demand. It is certainly too early to declare a turning point. Nevertheless, it appears, albeit tentatively, and is expected to gain momentum by 2027 with the introduction of the Chinese Sustainability Disclosure Act.

The effects take time to manifest when a situation must first be reassessed, as in the case of combating climate change. This quickly leads to demands for government intervention and regulation, particularly in Europe today. Governments set goals and prescribe the path to reach them, threatening sanctions if deviations occur. The result is often the opposite of what was intended: it takes longer, costs more for all parties involved, and quickly loses public support which is essential for the success of such initiatives. Above all, this mistrust of the economy leads to an explosive increase in bureaucracy and administrative costs. Prince Michael von und zu Liechtenstein, founder of the information and opinion platform "Geopolitical Intelligence Service," told Diplomatic World: "While more and more skilled workers are missing in the economy, employment in administration continues to rise – in areas where no social product is created." He considers this a clear misstep, and the numbers support him: in 2024, Germany lost 100,000 jobs in the economy, while exactly the same number were newly created in administration.

It seems we are gradually approaching a critical threshold. Over the last 20 years, bureaucratization has gained considerable momentum. In Germany, craftsmen now report that the ratio of on-site work to administrative tasks has reversed, forcing them to hire additional staff just to maintain revenue. This is an alarming development in times of an alleged skilled labor shortage.

Deep Impact

Everywhere one looks, the same picture emerges: in healthcare, particularly among general practitioners; in the housing market, where regulations make construction and even renting overly complex, expensive, and risky. The list could be extended indefinitely, and all of this hampers the economy. Blaming the European Union alone falls short. Brussels indeed intervenes in far too many areas, and the coordination with member states, which want to safeguard their individual interests, delays processes and increases the intricacy of directives. About 60 percent of national regulations in EU countries originate from Brussels.

Germany, however, generally interprets the Commission's guidelines more strictly than its neighbors. For example, the General Data Protection Regulation is fundamentally established, but its implementation was left to individual member states. While most countries have relatively moderate rules, Germany's regulations are particularly stringent. A similar pattern is emerging with the final design of the Working Hours Recording Act.

The harmful effects of over-bureaucratization have long been recognized in Europe and even reached the recent German "Ampel Koalition" (traffic light coalition). Yet the measures announced by the Ministry of Economic Affairs appear half-hearted, even under Chancellor Merz. The European Parliament and Commission have long debated implementing the principle that four old regulations should be repealed for every new one. This initiative is stuck in the thicket of bureaucracy. At least the European Commission has backed down now a bit in some areas for example with the General Data Protection Regulation. However, to proclaim this as a genuine change of course, as some experts did right after the decision was announced, is clearly still premature. Ideological legacies would need to be cut, battles long fought notwithstanding. Likely, our economy must suffer much more before real change becomes possible. These are not encouraging prospects in the short to medium term.

The impact on the economy is particularly severe: supply chain laws, accessibility laws, anti-money laundering regulations; all meticulously documented by companies for potential inspections, sometimes under the threat of enormous fines. This consumes enormous manpower. Large corporations can handle it, but for medium-sized companies, it becomes problematic: they may have to decline orders because compliance costs exceed profits, or the risks are simply too great. Medium-sized enterprises, with their limited personnel, suffer most, as they lack resources for proper implementation, whereas large corporations can better absorb these burdens.

Deep Impact

Regarding new technologies, data protection, and similar areas, the bureaucratic jungle is especially dense. How is innovation to flourish here? In Germany, with its unique nuclear phase-out in energy supply, high energy costs are an additional strain on the economy, particularly as energy consumption continues to rise. In the Frankfurt metropolitan area, Amazon has announced roughly seven billion euros in new data centers. Other platform operators have similar plans. Yet implementation is delayed by several years due to energy scarcity. Even though Frankfurt, as a European hub, is particularly affected, similar phenomena are observed worldwide, driven by tech network companies of various kinds. Energy prices are therefore a crucial competitive factor and a major contributor to rising inflation.

The opposite of inflation is deflation. The handling of the latter deserves critical analysis. Perhaps I stand alone here: I frankly do not understand why central banks treat deflation as a specter and sell moderate inflation as "stability." Falling prices would shift investments into the future, slowing the economy, they argue. This seems exaggerated: one or two years of mild deflation, with uncertain reversal, could arguably even incentivize investment. The real problem seems to be the over-indebtedness of the so-called "public sector," i.e., governments. Inflation indirectly reduces debt, whereas in the opposite scenario it stagnates or increases. This also applies to companies, though they are generally more adaptable. For states, however, it is a major challenge. Is government debt a central reason why inflation is normalized while deflation is demonized? We should be more aware and vocal whenever state debt becomes an issue. This compromises monetary stability and can be instrumentally exploited. Many established economists, in my view, argue too much in governments' favor, unintentionally creating another innovation barrier with implications for startup financing, among other things. This discussion needs to happen, and I look forward to it.

Startups also deserve mention. In terms of innovation, these small, agile "boats" are particularly important. With the right conditions, they can act faster and implement new ideas more readily than large, cumbersome incumbents. However, bureaucracy makes both starting up and surviving particularly difficult, as they often lack the resources to fully comply with rules. Large corporations manage much more easily. Financing is another challenge. Europe struggles to create globally competitive structures, preventing German innovations from reaching the market. The classic example is the MP3 standard, developed by Germany's Fraunhofer Institute, but not implemented domestically. It was sold cheaply to Apple, which turned it into a global success story. That's a pattern repeated ever since.

Deep Impact

"We must finally implement our own innovations at home", European Investment Bank Vice President Nicola Beer and Monika Hohlmeier, chair of the European Parliament Budget Committee, agreed during a fireplace discussion during the Neudrossenfeld European Days in May 2025. Nicola Beer noted that markets in India, China, and the USA are more coherent and easier for investors to navigate. The EU, with 27 member states, is disadvantaged, a fragmentation mitigated in part by the Investment Bank. "€33 trillion in savings lie in European accounts," she emphasized, highlighting that private capital could and should support future investment initiatives. The first European Innovation Fund for new technologies has already been successfully implemented, benefiting both Europe as an innovation hub and investors from all societal strata.

European authorities also need to rethink their approach. Monika Hohlmeier recalled that a few years ago, the antitrust authority did not recognize adequate value in the massive data flows of major international platforms. At least the problems are recognized, and concepts, albeit delayed, have been developed. Yet implementation remains trapped in ossified structures. This was the European perspective; similar issues manifest differently in other markets.

Deep Impact

In a nutshell: fundamental reform of our increasingly "late-capitalist" economic system seems overdue. Terms like "deindustrialization" or "degrowth," or even "economy for the common good," are proposed as alternatives. The pyramid scheme of perpetual GDP growth has many negative consequences, as socially harmful factors count positively as long as they generate growth.

In my first book on this topic, *Pandemias Box*, I gave an example that remains highly relevant. To grow in a saturated textile market, clothing companies shorten the intervals between new collections and special offers. The domestic market suffers from declining quality and prices, and an overflow of barely worn used clothing has collapsed the secondary market. Boston Consulting calculated that in 2025, this generates €130 billion worth of material waste annually. Most cannot be recycled due to blended fabrics and ends up in landfills. Vast quantities of cheap used clothing are exported to poorer countries, such as in Africa, undermining local markets. On the raw material side, demand for cotton continually rises. The Aral Sea, once the world's second-largest freshwater reservoir, is nearly dry, as water from inflowing rivers is diverted to irrigate cotton fields.

These processes fragment into countless transactions, all recorded in GDP, increasing its value. GDP growth reflects transaction volume, not quality, economic or social. GDP can grow even as the population's well-being declines. Continuous growth is impossible; pyramid schemes collapse sooner or later. Why should the global economy differ? Serious voices predict a decline in GDP's relevance in coming years: OpenAI investor Vinod Khosla forecasted in 2023 that AI will deflate the economy over the next 25 years, reducing GDP's significance. From a classical economics standpoint, this may seem alarming, especially as it comes from an investor, not a leftist ideologue.

What alternative exists to GDP? It measures economic performance, not well-being. Economy should serve humans, this is a value I internalized as a child. Today, we are far from that. Profit maximization trumps job security. "Shareholder value" is the late-capitalist golden calf. Companies once satisfied with 5% profit margins now aim for 10–20%, often with products in already saturated markets. Efficiency gains from technology should ease labor, but staff reductions have the opposite effect: remaining employees face increased workloads. Without solutions, society itself becomes destabilized. We may face unrest reminiscent of the late 19th–early 20th century Industrial Revolution but potentially more severe. Calls for rethinking are growing, often strengthening failed leftist ideologies, yet old recipes enjoy a resurgence.

Deep Impact

Is the much-touted call for deindustrialization an inevitable consequence of technological progress? Studies increasingly suggest yes: rising productivity, free cloud tools enabling complex DIY products and services, all drive unsustainable price declines. Even energy is projected to be abundant. While we cannot know if these predictions will materialize, adjustment processes will be difficult, especially for traditional industries. This process has begun and shows no end, potentially leading to a new economic system.

I consider it unlikely that a "economy for the common good" will ultimately prevail. Its appeal is obvious: only companies contributing to the common good may freely operate. Yet it effectively leads to a planned economy, with administration and tax authorities monitoring compliance. This demands bureaucracy, opens the door to lobbying and special interests, and constrains both economy and society, likely achieving the opposite of the concept's intent.

Continuing to focus solely on growth risks hitting hard limits. I noted this in my first book: in 1972, MIT scientists published the Club of Rome's Limits of Growth, warning that civilization could collapse in the 21st century due to overexploitation of planetary resources. Initially ridiculed, the study has since gained serious confirmation from Gaya Herrington of KPMG in a 2020 study, published in the Yale Journal of Industrial Ecology. It concludes that the current "business-as-usual" course may lead to declining economic growth within a decade and, in a worst-case scenario, societal collapse by around 2040.

Even if I avoid predicting societal collapse, a major economic and financial crisis seems plausible soon, with social repercussions. We must seriously reconsider tomorrow's economic system and expand the concept of sustainability beyond climate. Deep tech, especially AI, is already reshaping the landscape.

In my view, there is only one way forward: capitalism must return to quality and durability, abandoning growth at any cost. Growth will continue in innovation and sustainable investments. Let us now turn to society, which is naturally deeply affected by all of this.

··· on society

Innovation pessimists might even find something positive in this, as it could slow down the pace of innovation and, with it, its impact on society. "Degrowth" or deindustrialization are fairly common terms in today's debate, as is the refusal of innovation. We have already discussed the prospects of permanently refusing technological progress. They are virtually zero! We have already touched on "degrowth" in terms of the economy and will revisit it in connection with climate change. It certainly has a social component as well, with direct effects on wealth and well-being. Both are all too often overlooked in current discussions. The topic fits seamlessly into the debate within our society, which, controversially and pointedly, can be seen as an attack on a free society. The factions are increasingly irreconcilable.

Here is an example from the other side, taken from the German news portal n-tv on June 10, 2025: "Slovak Prime Minister Robert Fico has caused a stir by praising the economic 'efficiency' of authoritarian states and criticizing European democracy. During a visit to Uzbekistan, the left-nationalist politician told Slovak journalists: 'It seems to me more and more that in Europe we need to consider reforming the political system based on free democratic elections so that we remain competitive.' Countries like Uzbekistan, as well as China and Vietnam, are economically more efficient because they can act more decisively.

Only after explicit questions from journalists did Fico clarify that he was not advocating the abolition of democracy. According to him, one should be 'inspired' by systems structured differently: Fico advocated reducing the number of political parties involved in state decisions to speed up decision-making processes: 'If you have a hundred political parties, you cannot compete,' he said. 'If you have a government made up of four political entities, you cannot compete.'"

This opinion has now become quite widespread. Autocratic societies appear economically more efficient. Examples around the globe, from Singapore to Dubai and, of course, the countries mentioned by Fico, confirm this. Even the former British crown colony, now part of China, Hong Kong, is said to be further developed along these lines, serving China as an international business hub. This represents a new form of autocracy. Everyone is free to do as they please, as long as they do not question public order or criticize political leadership. It conveys a subjective sense of freedom. The cage as a large open-air enclosure is no longer perceived as such. The less attractive Western countries appear and the more they lose their old role as a model both economically and in terms of freedom, the more superior this model seems.

However, one should not be deceived by the disadvantages. Any alleged misconduct can be immediately sanctioned, nullifying this subjective feeling of freedom for individuals or certain groups. There is also no guarantee that this seemingly successful model will persist in the future. Autocrats, especially when they consider themselves untouchable or, conversely, feel their power threatened, tend to make abrupt shifts. Nothing can stop or even slow this down.

Dubai, for instance, has positioned itself as a global innovation hub. Low taxes and huge state investment funds promise a veritable startup paradise. Gordon Einstein, a celebrated crypto lawyer who deliberately settled in the emirate several years ago, says Dubai was a revelation for him, both in terms of the talent it attracts and the atmosphere, which creates an ideal climate for "doers." That is the snapshot.

Schaik Muhammad bin Rashid Al Maktum, the visionary current ruler of the desert state on the Persian Gulf, who has significantly shaped the country's success, is credited with saying that his grandfather rode camels while he could afford luxury limousines. His grandchildren will ride camels again. The point he wants to make is that children eventually only consume what previous generations have built. We know this from family businesses, which are often passed down prosperously, only to wither, be squandered as shadows of themselves, or even go bankrupt.

The latter can also happen to countries if they respond autocratically, and it can happen quite quickly. But democracies can also be affected by such developments. Wealth is consumed, infrastructure decays… This will be all too familiar to many.

Deep Impact

However, in democracies, this process is slower due to the separation of powers, as different interest groups have a say. On the one hand, this slows the pace of innovation, but at the same time offers a chance to counteract it – at least in theory. Fico is therefore partly right when he insists that our democracies need reform. The question is only: in what way? I spoke with a renowned human rights expert, who warned of "a frightening trend toward autocratism worldwide." She is not alone in this impression.

Yes, our democracies must be reformed, regain their responsiveness to new developments, and return to what makes them strong. This will also restore their stability, which Western societies are currently far from achieving. On the contrary, they are divided, not only in the U.S., where the MAGA camp and more liberal forces face each other irreconcilably. Although the understanding of the term "liberal" in the U.S. is very different from in Europe. In the U.S., it is often used in the sense of "communist."

But not only there; everywhere, the radical spectra to the left and right of the so-called "center" are growing stronger. In attempting to defend the liberal system, people too often follow the first impulse and resort, albeit often indirectly, to means that actually belong in the radical camps. Indirect censorship, i.e., restriction of freedom of speech, for example: Only what aligns with my opinion may be expressed freely. Everything else is not forbidden but is at least socially sanctioned. In a somewhat delayed effect, such a supposed majority dictatorship mirrors its autocratic counterpart. The term "cancel culture" is an example. The opposite is achieved: society becomes further divided, and the camps grow increasingly irreconcilable.

On closer inspection, the reasons are obvious. Citizens are unsettled because society is changing very quickly. This has to do, though not exclusively, with technological progress. People no longer feel at home in their familiar environment. The world of work is changing; many feel left behind or increasingly fear they cannot keep up, risking their jobs and thus their livelihoods. This quickly leads to nostalgic romanticism. Everything seemed simpler, safer in the past. The internet promised to make everything easier. Today we see the exact opposite, at least here in Europe. Technological progress is overregulated, and apart from legitimate safety concerns, it becomes difficult not only for providers but also for users to navigate offerings. Simple is a thing of the past. Humans, in addition to all other problems, feel abandoned.

As if that were not threatening enough, regulation interferes even more extensively. Traditional family structures are increasingly, and seemingly rapidly, being dismantled to strengthen individual rights. At the same time, governmental distrust increasingly constrains individual freedom of action. In the 1980s, during my university years in Germany, I could start a theater project in an abandoned hall with scrap furniture without problems. Today, fire departments, regulatory offices, etc., are involved and cause such socially valuable initiatives to fail from the outset.

Society seems increasingly complex, but also more selfish. Traditional parties ignored the warning signs for too long. Not only in Germany, "non-voters" had long been the largest group. Occasionally, warnings were heard that this could become a problem. They were ignored. When this group in Germany became involved in the newly founded party "Alternative für Deutschland," AfD, initially a mainly economically liberal party, it was immediately labeled as far-right. The reasons for voters' choices were only superficially examined.

Instead of slowing the pace in certain areas or fundamentally reconsidering one's own position and making real compromises, ideologies clashed with predictable results: increasing societal division and strengthening of radical fringes on both sides.

Meanwhile, right-nationalist parties like the allegedly or actually radicalized AfD in Germany are gaining support, making them impossible to ignore. If a party like the AfD, with up to 26%

of voters according to polls at the end of 2025, is ignored, it is a severe burden on a society that wants to be understood as a democracy not only in name.

That was only a view of Germany. Other countries have similar situations, albeit with different formal paths. Fundamentally, the effects are very similar. The reason seems to be that the ability to make genuine compromises has been lost. Even in the so-called political center, it exists only in a limited way, which many citizens perceive as increasingly left-leaning. This is typical bubble formation, a bubble not only due to social media channels feeding users only content that reinforces their views, even if based on misconceptions. Unfortunately, many critics of this bubble do not realize they are also trapped in one. Which brings us back to the effects of technology on our society.

When looking at societal risks from technological developments, we must also consider digital, now AI-driven, "workflows," i.e., automated processes. I now see them as a real danger to our society. We have all experienced during online shopping or other online transactions that the process simply stalls. Often, this is because our case is an exception not accounted for in the workflow. Not long ago, you would call a hotline and a trained human employee would promptly and effectively help you. Today, you will most likely communicate with a chatbot trained specifically on the workflow where you got stuck. The bot will try to detect your "error" and will therefore find no solution. With bad luck, communication is simply terminated; with some luck, you reach a human employee, who, in most cases today, is not properly trained. They will also try to identify the "user error." Thus, we repeat the same fruitless procedure. If we still do not give up or are thrown out of the queue again, we might finally reach a trained specialist who immediately understands and solves the problem in minutes. But reaching this point - if we have not already given up – wastes a lot of our time.

The problem is that these digital workflows cover many, but not all, options. Of course, a workflow could be programmed to be more flexible and responsive. With AI, this is possible but requires much more complex software and enormous additional data streams and costs. Behind the costs and effort, another, possibly much more powerful reason is hidden: the company or authority has no interest in changing their workflows; they want us to adapt our habits to it. As a result, we lose individuality. We act less complexly, and in the end, everyone acts alike. This will remain a problem even with AI. It makes workflows more flexible, but it can also be used to make us all more "uniform" to fit into a mold. To avoid this, we must actively resist being subsumed by workflows and defend our individuality against chatbot diktats.

Deep Impact

These were just some examples of the causes of the ever-deepening divides in our society. Bridging these gaps is becoming increasingly difficult but also more necessary. This is evident in another global issue: sustainability and climate protection.

··· on climate and sustainability

Even when looking at the economy and society, we have seen that climate and sustainability will be defining issues in the coming years, if not decades. New technologies will help us become sustainable; the path of waste is tracked to feed the circular economy as completely as possible, and AI controls our heating systems for maximum efficiency. Yet the term "sustainability" is narrowly defined and subordinated solely to climate goals. I do not believe this is the ultimate wisdom. All these applications have a major disadvantage: they make everything much more complex. What is needed is exactly the opposite. Many people want clearer, simpler structures. Deep tech can support us here as well.

The problem is that the topic has been ideologically captured. Broad societal support is fundamentally needed – a "we can do it," as repeatedly propagated by different protagonists worldwide. Focusing on the solution of a single ideology calls this into question.

If a party criticizes the European "combustion engine ban" and calls for technological openness, this is immediately equated with successful lobbying, in this case by the automotive industry, and condemned as a step backward. It is not about extending the life of obsolete fossil fuels. If combustion technology is categorically excluded, further research in this area stops; that is, alternative sustainable fuels that could realistically enter the market and complement climate-neutral drive technologies, thus achieving climate neutrality faster, are no longer developed, at least not in this market. Progress in innovation always occurs where there is competition. Exclude competition by focusing on a single technology, and the opposite happens: progress slows.

Which brings us to general technological skepticism. Distrust of new technologies and attempts to regulate them in advance often slow or even prevent the development of new sustainable approaches. This also slows the achievement of climate goals. And here we are again, directly at the influence of ideologies on progress, which can even sabotage achieving their own goals, not necessarily, but with very high probability.

Deep Impact

When a problem like climate change is recognized, it always takes time for everything to align on the solution. In the meantime, we often see blind activism. People want to act immediately because they cannot wait, but they do not know what to do. They tinker with symptoms that do not fundamentally solve the problem. Meanwhile, little-noticed basic research in laboratories continues, whose results are transformed by clever entrepreneurs into marketable products, which, of course, take time to attract initial investors. Slowly, these products become established and recognized until they finally generate their own momentum. That is the point where we approach real solutions.

We are currently in this intermediate phase. Some products have been developed or are being introduced. Initial investors are on board. Of course, there are always those trying to profit from trends with pseudo-products, such as "greenwashing" in the climate sector, slowing the sector overall. The image suffers and delays the success of sustainable approaches. But in the long run, they cannot be stopped.

At the same time, we are still confronted with activists, but also with skeptics who gain influence as a reaction to blind activism, thus not only slowing down nonsensical and unnecessary initiatives but also delaying meaningful solutions. This includes the EU regulation that came into effect on September 30, 2025, intended to prevent further deforestation. This means that not only producers and importers of certain products like coffee, beef, rubber, or wood must prove the exact origin via GPS data of the cultivation area, but it also applies to processed products. This again involves immense bureaucratic effort, which helps the climate little but significantly increases the cost of products. The intention is certainly commendable, but the approach is entirely misguided. Yet it is also an example of how new technological possibilities, in this case, geotracking via GPS, can be used counterproductively despite good intentions. Unfortunately, this example is not unique. As a result, the economy shrinks, and CO_2 emissions decrease. This is not the merit of climate policy, even if activists welcome this deindustrialization. It destroys wealth and destabilizes society. Alternative solutions are therefore necessary – and possible.

Climate protection is an evolutionary process, one that also faces setbacks. It takes time, a lot of time, and thus patience. With patience, some setbacks can also be prevented or at least mitigated. The question is: will we still be fast enough? Some scenarios are dramatic and contradict the demand for patience. I would not dramatize the answer but still urge prudence. Patience, distinct from negligence, accelerates evolution; activism delays it.

Deep Impact

As we noted at the beginning of this book, in our overview of Deep Tech trends and developments, some technologies could have a lasting impact on the global climate. The spectrum ranges from CO_2 prevention to the removal of already emitted CO_2 from the atmosphere. These technological approaches are sometimes expensive but offer other added value. For example, solidified carbon dioxide can be used as a building material. Another approach currently being tested for market readiness is passive cooling in data centers, which will continue to have the highest energy demand in the future. Also, carbon trading, if used the proper way and not for "greenwashing", creates sustainable financial instruments that can help reduce the CO_2 footprint. The list of examples could go on indefinitely, and some may turn out to be real game-changers.

The "circular economy" is a model for increased climate protection and sustainability; "natural capital" and trading CO_2 certificates are others. Circular economy means there should be no waste. Everything should be recycled and reused for new products. The concept is not new. It has repeatedly appeared during times of resource scarcity, such as during wars. Today, it is seen as one of the most important tools for reducing CO_2 emissions and conserving resources. However, there are still limitations. If you want to recycle old electrical cables, there is a simple method: heat the cables, burn off the plastic coating, then melt the metal, with the ash floating on the surface. After skimming it off, the metal is poured into molds and cooled. At the end, you have pure copper bars that can be reused industrially.

In principle, the process is fairly simple. For this reason, scrap metal has long been recycled, albeit not consistently in the past. The process itself is not very environmentally friendly or sustainable. Even when using renewable energy for the smelting process, there are emissions and ash, as the plastic in the cable sheath cannot yet be recycled. Of course, recycling processes have generally improved and will continue to do so. But there is still residual waste that cannot be reused. Many recycled materials can only be used for certain products. But what is the market for garden benches made from recycled plastic or for river bridges from old wind turbine blades?

It is still difficult to capture all waste, so much is lost and also pollutes the environment in illegal landfills. Blockchain technology promises applications to track waste and keep its materials in circulation as completely as possible. This will facilitate future, nearly complete waste utilization, even if such a tracking system cannot be implemented everywhere and is highly complex in practice.

The circular economy is certainly important, but what makes me skeptical is that the focus on it allows us to continue our growth model unchanged, at least for a longer time. Returning to our example from the clothing industry: we can continue to increase marketing expenses to encourage people to buy more of what they will never wear, assuming it will be recycled later. The environmental burden remains high. This is greenwashing. The problem becomes less visible but is not solved. Waste avoidance will therefore remain a major issue.

We must also return to a more rational use of resources by increasing product quality and repairing them when necessary. Less is sometimes more. Even in expensive, supposedly high-quality products and spare parts, solid components were sometimes replaced with plastic parts, integrated in such a way that they cannot be accessed without destroying the housing. It is known that the plastic part will break in two years, whereas the more expensive metal or ceramic part would have lasted at least ten years. But the consumer is encouraged to consume regardless of the cost. That is not sustainable! We need both a circular economy and a more responsible approach to resources!

A similar standard must be applied to other concepts. "Natural capital" is said to create a market for sustainable services and products. Only when money can be earned with nature do we recognize its value, says Ralph Chami, former Deputy Director of the International Monetary Fund and now a pioneer in this field. Forests, i.e., wood, have long had economic value but were often simply exploited because they seemed inexhaustible. In primeval forests, such as in the Amazon, this still happens, whereas in developed countries like Finland, reforestation is at least monitored during wood production. The rest of nature had no commercial value; in other words, its value was zero. "But that is also a value," Chami notes.

Nature has a value potential, as we are gradually realizing. Take mangroves and seagrass, for example. Mangroves have long been considered useful for protecting coasts from erosion or for land reclamation in coastal waters. Yet mangrove forests are now considered critically endangered. Seagrass is also threatened. Until recently, it had no known utility. This is changing. Both store CO_2 significantly better than trees, mangroves five times, seagrass even 30 to 50 times more. Seagrass could store 45% of the CO_2 already present in our atmosphere if all areas suitable for its growth were restored, says Chami. This represents enormous economic potential for countries like the Bermuda Islands.

CO_2 has a price per ton. Companies that cannot immediately reduce their emissions can purchase emission rights, for example from Bermuda, which is establishing large seagrass farms. It is a win-win situation: Bermuda creates a new income source, and the company can continue producing its products a bit longer without additional high investment. There are also other models for monetizing wildlife protection. This allows companies to generate returns with a healthy environment.

However, there are also downsides. Carbon trading is considered one of the most important tools for rapidly and significantly reducing the global CO_2 footprint. This is true as a catalyst for change. In the long term, however, it could slow CO_2 reduction. Returning to our example: Bermuda would have no interest in a company that maintains its emission rights through purchasing CO_2 certificates investing quickly in its own CO_2 reduction. When speaking with advocates of this certificate trading, it always seems they consider it a universal tool for fighting CO_2. If applied too rigidly and as the only "truth," it can lead to a dead end. Already today we see healthy and bis diverse woodlands with high carbon storage potential being destroyed for renewable energy timber to be transformed into charcoal, unhealthy monocultures, for "carbon neutral" steel production. The farms funding comes from carbon certificates. This is greenwashing, creating more harm than good.

We need genuine technology neutrality. That is, we should not focus on specific technologies (or strategic approaches). Focusing on one technology makes other R&D approaches unattractive. They develop, at best, delayed, or in other countries that then "suddenly" overtake us. This leads to less competition, delayed solutions, and higher costs.

For example, the Australian company MCi Carbon could provide another solution to reduce CO_2 already in the atmosphere. Co-founder Marcus Daws is the laureate of the Wholistic World Innovation Trophy 2023, facilitated by DWI. MCi technology extracts CO_2 from the atmosphere and, through a chemical reaction with calcium or magnesium, converts it into stone, which can form the basis for building materials like cement or concrete. This simultaneously helps alleviate the shortage of construction sand, which is becoming a serious global problem. The process is expensive, certainly a disadvantage, but it also offers a solution for another problem. There are other similar projects. All contribute to making our planet more sustainable.

We must monitor all of this very carefully. Mistakes are easily made because not all facts are usually known when a decision must be made. I am sure that some developments we are proud of today will give us headaches in the future. The focus on renewable energy seen in some countries has that potential. For example, future generations may shake their heads at today's wind turbine hype. Huge components are shipped from China to Europe and placed on massive foundations after large-scale clearing of construction sites and access roads, especially on forested hilltops. Disposal at the end of life is also unresolved, as fiberglass is currently difficult to recycle. Considering all this, the final balance may not be as favorable as wind energy advocates believe today.

Unfortunately, mistakes can never be fully avoided, as the complete picture only becomes visible with some distance. What will happen with the "legacy" of wind energy, if fusion energy becomes available in the foreseeable future, is just as uncertain today as whether this alternative will arrive that quickly. Ultimately, the new energy form could also have negative aspects we are not yet aware of. However, we must stop ignoring warning voices simply because they do not fit our ideology.

... on national and international relations

Even today, we see the impact of deep tech on relations between countries. Over the past decades, the role of diplomacy has already changed significantly, for example, the role of an ambassador. This is also due to new communication channels that allow heads of state or ministers to exchange information much more directly and quickly. As a result, ambassadors focus much more on economic networking, essentially on economic promotion. This is already a significant change. Unsurprisingly, this will continue to accelerate.

I again asked AI, in this case, ChatGPT-5, about the potential for change over the next ten years. Here is what I find a very plausible assessment: "Deep tech will profoundly transform diplomacy over the next ten years. The effects can be observed across several dimensions: distribution of power, negotiation methods, security policy, and norm-setting.

Shift of geopolitical power centers: States that master deep-tech capabilities will increasingly gain strategic advantages. Similar to the nuclear age, technological breakthroughs in quantum communication or AI could reorder the global balance of power. Technological alliances, such as between research-strong democracies or between authoritarian states, will become central pillars of foreign policy. This could complement or even replace traditional military alliances.

Transformation of diplomatic tools: AI-supported data analysis will enable diplomats to monitor social media sentiments, economic trends, and security-relevant developments almost in real time. Simulations of complex negotiation scenarios could drastically improve preparation for international summits. At the same time, the risk will increase that states use AI-driven disinformation as a diplomatic lever, which complicates the basis of trust between partners.

New forms of security diplomacy: Quantum encryption could revolutionize secure communication between governments, while next-generation cyberweapons will become a constant topic of negotiation. "Tech arms control", agreements regulating AI weapon systems, autonomous drones, or synthetic biology, will be as important as traditional disarmament negotiations.

Global norms and values: Deep tech raises ethical questions: Should autonomous weapon systems be banned? How far should genetic manipulation go? Who controls global AI infrastructure? Diplomacy will increasingly become a platform for technology regulation, with conflicts between differing value systems, such as data protection orientation in Europe versus data-driven models in other regions becoming more pronounced.

Multiplication of non-state actors: Large tech corporations, research institutions, and even individual innovators will become diplomatically more relevant. They can be both partners and geopolitical influencers, forcing traditional foreign ministries to adjust their strategies. "Corporate diplomacy" and public-private alliances will gain importance."

Deep Impact

The conclusion is unsurprising: "Deep tech will make diplomacy faster, more data-driven, and technology-centered over the next ten years, but also more vulnerable to distrust, manipulation, and unregulated risks. Those who combine technological leadership with diplomatic skill will exert disproportionate influence on the international order. States that invest in research while also shaping global norms will hold the best cards in the geopolitical game of the future."

Much of what ChatGPT extracts in its web analysis is already visible today, at least in part, and in some cases even in a more advanced stage. Did you notice anything while reading the previous chapters? It feels like China is now ahead everywhere. Much of the technological progress now comes directly from there. And developments from the US or Europe are often quickly surpassed by reports from the People's Republic, sometimes significantly. This initially subjective impression can also be verified statistically.

"Visual Capitalist," a Canadian media platform for data visualization, shows the development of patent applications in AI between 2010 and 2025: Initially, the US was clearly in the lead with 40 percent in the global ranking. Just behind was Asia excluding China at 37 percent. Europe, with a share of just under nine percent, was already far behind. Today, the picture has changed dramatically. China, with less than 14 percent in 2010, now accounts for nearly 70 percent of AI-relevant patent applications. The US and the rest of Asia are now only at around 14 and 13 percent, respectively. Europe's share has even dropped to below three percent. In other areas of innovation, a similar trend is emerging.

The reasons are easy to understand: China has comprehensively analyzed future markets and trends and strategically occupies all key positions to consistently advance investment and research. The centrally organized state facilitates this approach. Not only public universities and research institutes are included, businesses are also massively subsidized. Already in the early 1990s, voices were heard predicting the West's decline and the rise of an Asian era. Over the years, this impression has solidified. However, it would be more accurate to speak of a Chinese era. China has already secured a clear lead. In many high-tech areas, the US still maintains a leadership role, but it is eroding quickly. The extent to which US government measures can reverse or at least slow this remains to be seen.

However, there are also risks that we can only partially assess from the outside. The Politburo of the People's Republic not only heavily subsidizes and controls, but also closely manages the information released externally. For several years now, we have observed a progressing market saturation with a noticeably slowing GDP growth. We also see a massive bubble forming in the national real estate sector, which has already caused some disruptions. Unemployment, especially among young people, is at a worrying level. All of this could lead to shocks with global impacts.

The reason for China's long-standing dynamic development lay, besides a high degree of discipline among its population, which until recently was the largest in the world, and thus the largest homogeneous domestic market. The political system, which calls itself communist, seemed for a time to open toward the West. Meanwhile, the Chinese leadership has returned to an autocratic structure with an undisputed chairman.

To better assess the situation, we should become familiar with Confucian thought, which now again exerts a strong influence on Chinese society and leadership. The Confucian concept of freedom is very different from the Western one, which is itself not homogeneous. We have already examined the differences between Europe and the U.S. in connection with ethics and law. In China, the concept is even more distinct.

In Confucianism, freedom is understood as moral self-cultivation within social relationships. A person realizes their freedom by fulfilling their role in family and society in accordance with the principles of Li, ritual order, and Ren, translated as humanity. Freedom here does not mean independence from social norms, but harmonious living in accordance with them. The Junzi, or "noble person," is free because, through self-discipline, education, and virtue, they achieve morally autonomous action, not despite, but through, their social embeddedness.

In the European, especially liberal, understanding, such as John Locke or Immanuel Kant, freedom is defined as the absence of external coercion or as the individual's autonomy to make self-determined rational decisions. Social order should protect individual freedom, not shape it. Here, the individual is often central, setting themselves apart from the state, society, or tradition.

The difference thus lies in the relationship between individual and community: In the Confucian tradition, the individual is largely defined by relationships and finds freedom in moral perfection within these structures. In the European tradition, freedom is more closely linked to individuality, choice, and self-determination, sometimes against societal constraints.

If we are surprised by the high acceptance of the Chinese "surveillance state," we find the difference precisely in this alternative thinking, which provides a possible key to respectful interaction.

Deep Impact

Another example: Confucius states that it is disorderly when princes rule, whereas the rule of the emperor represents order. This belief is deeply rooted in Chinese society and has kept the system stable over centuries, including the time after Mao's revolution to Xi today. At first glance, this seems surprising, but it is connected to the willingness to accept an absolute ruler. While old communist and post-communist leaders formally turned away from tradition (though the concentration of power under Mao remained and continued afterward), Xi today strengthens the old roots. The real existing communism fit very well into an absolute model of power, not only in China: Soviet leaders were often referred to as the "Red Czars."

This fits even more when looking at today's promoted Chinese-style communism. Confucius also states that the ruler should always freely give the people everything they desire, as long as it does not threaten his rule. This accurately describes the current situation of the capitalist-communist hybrid.

Another example is the concept of assimilation, deeply rooted in Chinese society. Confucius says everyone should have the opportunity to occupy any position in Chinese society, and intercultural marriages should be encouraged. In the long term, this leads to the complete assimilation of a cultural group, making it an integral part of Chinese society. This process made the

Han the largest ethnic group in the world. China was occupied several times throughout its long history. In the end, the occupiers were assimilated, not the other way around. The communist regime wanted to shorten the assimilation period by forcing people, for example in Tibet, into this process. Today, the same approach is applied to the Uighurs. Whether it is as effective as the longer path of patience is questionable.

Among the population, this approach seems generally well received. "Divide and rule" long ensured that living conditions steadily improved. Opportunities for development seemed limitless, as long as people stayed out of politics and did not criticize the regime. Even if this record is no longer flawless today, it continues to resonate, fueled repeatedly by success reports in science and technology. Additionally, the Western societal model now shows clear cracks in its formerly flawless image, weakening its role as a model.

What all these considerations show is that we must leave our Western perspective and understand this way of thinking to grasp how the system functions. This also includes understanding why the government is so well received despite all surveillance. In China, we are dealing with a completely different concept of freedom and a different self-understanding. Understanding these differences is the best way to find an answer to how we can position ourselves economically in competition with China. It also helps to understand the country's confident stance toward its neighbors and the US.

To complete the picture, another player in Asia must be considered. India has already slightly surpassed China in population and is developing enormous economic momentum. Since the country is still in an early stage of development, we can see the potential for a development trajectory similar to China. Additionally, there is growing interest in the US and Europe to reduce dependence on China. This gives India a trump card, which is being skillfully played. However, risks remain, such as the conflict with Pakistan over Kashmir, much of which is administered by India, complicating reliable forecasts.

Even if we assume India is the next tiger state we should be cautious not to downgrade China or assess its potential problems prematurely. China has repeatedly demonstrated remarkable resilience over its millennia-long history, an ability to adapt and reinvent itself from which the West could certainly learn. However, this long history also brings an almost unshakable self-confidence.

The goal is to restore its former stature and regional dominance after decades of weakness, a dominance that will undoubtedly also be pursued globally. The project of the New Silk Road has brought this goal significantly closer. Here, the West could have engaged China on equal footing. "It was certainly a mistake to leave the New Silk Road initiative entirely to China. We should have been much more involved. But the mistake is made, and we must live with it," says Prince Michael of Liechtenstein, in the previously cited interview with Diplomatic World.

Although much has been neglected in recent decades, this provides an opportunity to respond to the challenge from China: cooperatively rather than confrontationally, to take up initiatives and participate, so as not to leave the field entirely to the People's Republic. At the same time, we are not only part of these initiatives but can also influence them in our interest. This, of course, requires unity. And here, the situation currently looks rather grim, as shown, not least, by a look at the European Union.

Many already see the EU in a deep crisis: Euroskeptic and nationalist movements are gaining strength in many of the 27 countries and have either already assumed government responsibility or are on the verge of doing so. How deep this crisis runs remains to be seen. However, the situation should not be taken lightly. The reasons are obvious: long-overdue reforms have been neglected, while bureaucratic overreach continues to spread.

Deep Impact

The cause of both lies in the structure of the Union and its rapid growth after the fall of the Berlin Wall. The Lisbon process, which was intended to streamline the EU, fell victim to the eastern expansion. One must credit decision-makers that there was little they could do to oppose the rapid accession of former Warsaw Pact states, as the goal was to help them integrate internationally. The price, however, was a structure chosen by the EU founders under completely different conditions: fewer states, roughly similar economic levels, and similar cultural roots. Rapid growth and the resulting size make decision-making processes in the Union more complicated and time-consuming. This further stimulated bureaucratization.

That bureaucratic momentum has accelerated in the past decade is, in my opinion, linked to Brexit, the UK's exit from the EU. The British have traditionally taken a much more market-liberal stance than the continental countries, which naturally influenced European Commission decisions. This balancing function suddenly disappeared.

Additionally, the burden caused by EU bureaucracy and over-regulation varies by country. EU directives are "framework regulations" that must be adapted into each country's law. While some countries implement them more liberally, Germany tends to be more restrictive. This affects the pace of innovation in each country, which can vary significantly.

This raises the question of Europe's overall innovation pace and its adaptability to international challengers, such as China. At first glance, the record appears mixed. There are indeed digitalization pioneers in Europe, such as Estonia or Malta. What stands out are the smaller countries in the Union, which, due to their size, or rather smallness, are more agile. Germany's digitalization record, however, is rather poor. The same applies to the EU overall. Not least, Germany's federal system, essentially adopted for the EU but made even less flexible by the national independence of member states, slows development and reforms. "Too many cooks spoil the broth," as my grandmother used to say.

RESILIENCE

What Next?

We've examined many areas and gotten a sense of where the causes lie, and realized that even where we don't expect it, there are connections that need to be considered. We have seen which technological advances influence our lives even more strongly and quickly than previous developments ever did. And we have seen how they interconnect, how one technology makes another more efficient or accelerates its development. Inevitably, we've also caught a glimpse of the dark side: the risks that progress and its speed bring, as well as the benefits these technologies can offer us.

This brings us to the other side of development, because we have also seen the impact these technologies have on our entire society; effects whose full scope we cannot yet gauge, at best merely anticipate. All of this generates expectations and tempts clever minds to greedily try to gain unilateral advantages, while others either fail to recognize the potential at all or only do so very late, often plagued by fear.

And it is precisely from these fears and this greed that another problem arises, one that must be addressed: future pessimism,

What Next?

fear of progress, and fear that technologies might destroy our society. This leads to overreaction and overregulation.

Regulation always comes too late because implementation takes far too long. By the time it becomes law, the situation has already fundamentally changed. It cannot be contained. Why can't we get a grip on the situation, especially since, in the previous chapters, we have already derived actionable recommendations? Why do we struggle so much to act on such guidance? Are we really so powerless?

Some scientists see it that way. "I used to think the biggest environmental problems were biodiversity loss, ecosystem collapse, and climate change. I thought 30 years of good science could tackle these issues. I was wrong. The biggest environmental problems are selfishness, greed, and indifference. To deal with them, we need a cultural and spiritual shift. And we scientists don't know how to make that happen." This quote is from Gus Speth, Professor of Environmental Policy and Sustainable Development at Yale University. He was also chief advisor to the Environmental Commission under U.S. Presidents Jimmy Carter and Bill Clinton. I understand his frustration. Yet he is clearly speaking from the perspective of a natural scientist. Other scientific disciplines, such as neuroscience and behavioral research, at least offer some potential solutions here.

What Next?

The role of our mindset

The biggest problem regarding the acceptance of change remains humans themselves, or rather, their "mindset", a term describing a multi-layer condition. To find timely educational approaches for behavioral transformation, we must understand our brain and our mindset. Today, when we talk about mindset, we usually mean individual mindset, and it primarily concerns one thing: self-optimization. We are told we can achieve anything if we program our mindset correctly. On one hand, this is very egoistic and ignores responsibility to the whole. On the other hand, it ignores limitations and deeply ingrained patterns that cannot simply be overwritten like a computer program. We cannot work against them, but we can work with them. That is a very different approach.

At first glance, this investigation might seem to lead to deep despair. It deceives us into seeing alternative realities and misleads us into misjudgments everywhere. At least, this impression easily arises when looking at studies that attempt to explain our behavior and especially at the interpretations of their results. I will highlight a few examples relevant to our discussion, simplifying a highly complex subject. At first glance, the picture is simply discouraging and seems to support Speth's assessment.

What Next?

The study is not even new. In 1980, an experiment was conducted at Dartmouth University under the guidance of social psychologist Robert E. Kleck to understand our perception of "reality." The results shocked at the time. The reason is still understandable today: it suggests that we largely live in a dream world. Participants were given realistic-looking facial deformities, birthmarks and scars, and asked to walk on the street and report how they were treated. Just before leaving, the makeup artist came back to "correct" the marks. After their outing, the participants reported consistently experiencing subtle rejection from the people they encountered.

However, the makeup artist had removed the deformities entirely. The perception of rejection was entirely due to the participants' own expectations. I found a comment on this that I cannot attribute, but will quote anyway: "I immediately think of all the discussions about discrimination. How much of it is objective, and how much is subjective, because media at the moment nurture and legitimize hypersensitivity? Nothing flatters the ego more than a victim role, from which one can subtly be a perpetrator and accuse others."

The question, however, is even broader. What is reality? The study shows: the brain does not show us reality. It shows us what we expect. What we perceive in the outside world is subtly

What Next?

guided by what is already part of us, our experiences, expectations, and, not least, deep-seated fears. That's why two people walking the same street can perceive entirely different realities. We saw this during the COVID years. Depending on the trigger (fear of disease or fear of losing freedom), people experienced vastly different realities.

The problem is not subjectivity. The problem is that most people believe they are objective. The commenter answers the question, "Why can't people agree on simple facts anymore?": "Because most people don't see facts. They see fragments, and the brain combines them with predictions drawn from its own horizon of expectations."

He tries to capture the issue with some pathos: "Now scale that up. A planet full of nervous systems projecting their fears and ideals onto the world, each convinced it sees clearly, each emotionally certain its version of events is 'reality.' The people in the study were not lying. They did not invent their experience. Their pain was real. And that's the frightening part. You can suffer deeply over something that does not exist. It's not about dismissing that pain. It's about taking responsibility for your perception. Not feeling better or thinking positively, but learning to interrupt the hallucination."

What Next?

But stop! The commenter goes too far here. I quote him at length because it is an excellent example of how individual studies can lead even intelligent people to overly one-sided judgments. In fact, his assessment was even harsher than quoted here. Naturally, we do not all fundamentally live in a hallucination, although our experiences obviously condition our perception.

This means that if there is a basic mistrust, it is hard to counteract. For example, if a government loses trust in its measures, it is difficult and time-consuming to rebuild that trust, because even if something goes in the "right" direction, a "wrong" approach is still assumed. Conversely, a certain basic trust makes it hard to recognize even nonsensical behavior. One always looks first for an explanation that fits their familiar perception. This must be considered when implementing measures that require broad consensus.

I quote the comment also because it narrows focus to a single aspect of a complex reality. Other aspects are consistently omitted: cognitive dissonance or the Dunning-Kruger effect, for example.

We've all experienced this: we explain a matter in detail, only to realize the other person understood nothing, even though they act as if they did. Only keywords that fit their worldview are

What Next?

retained. I must admit I sometimes fall into this myself. This is called "cognitive dissonance," a psychological state of inner tension that arises when a person holds contradictory cognitions (thoughts, beliefs, attitudes, or values) simultaneously, or when their behavior contradicts their beliefs. The term was introduced in 1957 by American social psychologist Leon Festinger. People strive for internal consistency between thoughts and actions. When this consistency is disturbed, they feel discomfort, so-called dissonance.

To reduce this tension, people adjust their cognitions or behavior. They may change attitudes, justify behavior, or avoid information that reinforces the contradiction. A classic example is smoking: a person knows smoking is harmful but smokes anyway. To reduce dissonance, they may downplay the risk by saying for instance "We all have to die of something"; or overestimate control. "I can quit anytime" is typical for that. It can also mean ignoring or misinterpreting statements that don't fit one's worldview.

Cognitive dissonance is central to understanding human motivation, decision-making, and behavior change. It explains why people cling to beliefs despite clear facts or rationalize behavior. It is a fundamental mechanism for maintaining psychological

What Next?

stability and self-image. It shapes our comfort zone, the area where we feel at home and safe.

Another aspect, though not the only one, is the Dunning-Kruger effect. This cognitive bias occurs when people with limited knowledge or competence in a domain systematically overestimate their abilities, while competent individuals often underestimate their relative performance. David Dunning and Justin Kruger empirically described this in 1999 at Cornell University.

The cause is paradoxical: those with little knowledge often lack the metacognitive ability to recognize their own incompetence. In other words, it takes some knowledge to understand how much one does not know. Dunning and Kruger found that the worst performers rated their abilities highest, well above average. Conversely, top performers rated themselves as average, assuming others would perform similarly. This occurs in careers, political opinions, or daily life. People with little exposure to a subject may confidently hold false beliefs, a phenomenon especially visible on social media.

The Dunning-Kruger effect highlights the importance of self-reflection, feedback, and education to realistically assess one's abilities. With growing knowledge, self-assessment often improves. That's a process Dunning and Kruger call "competence-induced humility."

What Next?

Everything we've discussed falls under the general term "bias." This is one of the biggest obstacles to accepting innovations and adapting to changes across all levels of our increasingly challenged existence. Education and improved self-assessment are certainly relevant, though they do not always mitigate the effect. A lawyer or business graduate may apply learned skills within their limited reality, often successfully. Only at the top, when decisions determine the survival or collapse of a company, do these limitations become apparent.

Ideologies, in my observation, also exhibit Dunning-Kruger characteristics because they claim to possess the one salvific truth. For example, socialism's attempt to make society more humane and just via a dictatorship of the proletariat is doomed from the start: it simplistically constrains complex reality and ignores key factors. Humans seek simple answers; ideologies promise them but cannot deliver.

Here again, everything interconnects. Reality is multi-layered, and perception depends heavily on perspective. Our perception is finite; we can never see the whole picture. Thus, entirely different realities reach our brains. One sees a "9," the other a "6." Both are correct from their viewpoint. In China, historical dominance in East Asia shapes self-perception and leadership claims. In the West, a different slice of reality informs self-image. In a globalized world, we must learn to navigate these contradictions.

This is partly an educational issue. In less-educated populations, reflection is often absent; conspiracy theories flourish, though educated groups are not immune. Comfort zones are needed. Demographics matter: even recognizing a new technology, I need time to change familiar routines, while young people adapt quickly but only to innovations encountered in formative years. Later, routines harden. This was historically advantageous but is now a liability in rapidly changing times.

This explains why the capitalist system is so deeply ingrained and why system-change attempts have failed. Our mindset is geared toward competition and advantage, essentially "capitalist." Real-world socialism too had elites enjoying privileges while others lost interest due to lack of reward beyond security. It failed because it assumed an idealistic view of humans. The Chinese communist model now operates in name only, offering capitalist incentives to loyalists. The implication: global problems must be solved within the current system. Our ingrained mindset favors and demands a form of capitalism. But we can shape its expression to ensure everyone shares prosperity. Current redistribution proposals are far from this; they are bureaucratic, expensive, and clash with our mindset.

What Next?

Roadmap out of the trap

Solutions must work through incentives that make broad wealth distribution economically attractive. Many behavioral patterns and perceptions are ancient, rooted in thousands of years. To solve complex global problems, we must work with our mindset, not against it. Our mindset prioritizes short-term advantage over competitors, ensuring family survival and children's start in life. Long-term collective strategy is far harder.

The same applies to distrust of strangers. Individuals may be welcomed, but larger groups quickly trigger mistrust or rejection. Historically, food scarcity made this adaptive. Today, it still matters for migration strategies. Ignoring these fears is counterproductive.

What does this mean for society, legislation, and education? We must learn to think holistically and long-term. We cannot work against our mindset, only with it. I recall a conversation with Frankfurt neuroscientist Henning Beck. Long-term solutions require involving people so they perceive immediate benefits. We want to be rewarded!

Legislation should incentivize desired behavior rather than imposing prohibitions or costly measures. For example, to encourage citizens to adopt environmentally friendly heating systems, devices must be affordable and immediately reduce energy costs. Norway succeeded with heat pumps this way. Coercion through expensive regulation and higher taxes produces resistance.

This principle applies to new technologies: digitalization initially eased life, but overcomplicated regulations have hindered adoption. We must work with mindset, not against it, to build resilience to change.

Which system enables this? Our "late" capitalism with its sole focus on shareholder value, has long exceeded human scale. Yet other values are equally important. An economy prospers only in a stable, healthy environment. Historical profit expectations were moderate; today, they are extreme, achieved by cost-cutting rather than growth. Earlier industrialists treated employees as family, providing housing, healthcare, and education, and deferred profits in downturns. Some family-owned firms still operate this way. Perhaps this approach should be central.

For financial markets, purely profit-maximizing behavior destroys stability, harming profits themselves, like a Ponzi scheme. Even renaming "capitalism" as "market economy" could help depoliticize the term. Adhering to limits of growth could prevent catastrophic scenarios predicted in studies.

Thus, applied wisely, capitalism can alleviate many major problems rapidly. This requires a liberal regulatory framework; an "ordo-liberal" level playing field ensuring opportunity, limiting greed, and fostering adoption of innovation. Benefits must be real and immediate. Avoiding bureaucratic excess is critical; otherwise, good intentions backfire, restricting and stifling both public welfare and our individual innovation resilience.

Resilience – Are we ready?

If we want to achieve change, it can only happen in alignment with our mindset. Change is only possible collectively, when everyone – even in the short term – recognizes a benefit for themselves. The answer lies in education, and this brings me back to my statement at the very beginning of this book, not just for our children or for those who reject our democracies, but for society as a whole, including those who lead it. We talk a lot about "lifelong learning." That means we must adapt throughout our lives to ever-new platforms, tools, and ways of working.

But it means even more: we must accept the necessity of long-term thinking and action, instead of continuing to search for short-term solutions. We must adjust our comfort zones to new possibilities rather than clinging to familiar surroundings. Some people do this very quickly, but they are by no means the majority. Unfortunately, this also applies to the decision-makers in our society. Too often, they are stuck in analog thinking, and they must accelerate adaptation processes. These processes need to happen much faster across society, even if they are perceived as "against nature." It is a lengthy process.

Of course, the younger generation, the so-called digital natives, who grew up with the technology, adapts faster. Naturally, there are differences; some are more rigid in their thinking than others, even from youth, with potential reinforcement as they age. Often this is just a cliché, but one deeply entrenched, for example among HR managers. This is something we also need to address.

We must work across generations, but not everyone sees it that way; some believe they can resist it. A graphic designer I know says he will fight AI all his life. His struggle is doomed to fail. The pattern is long known: first, we tend to ignore change, then we passively reject it, and finally, we fight it. That is the final stage. We do not succeed this way. We tend to ignore or even reject scientific facts to avoid leaving our comfort zones.

We must learn to accept that selfishness and intolerance ultimately harm every one of us. This is not a new insight, but it has never been more important to heed it. If we continue as we are, we will lose the battle for the future. The problem is that through our rejection, we often at least locally delay change. On the one hand, this allows for a harmonious process of gradual adaptation. But if it becomes a general societal mood, local standstills occur, and we soon fall behind faster-moving societies. Yet the resistors stick to their agenda. They do not realize that by resisting, they are undermining not only society but above all themselves.

Ultimately, this means we need a completely different definition of education. I would call it "holistic education," as a working title. I am talking about an education system made up of various interconnected modules, from early childhood to lifelong learning, which focuses not only on practical skills but also on preparing our ways of thinking for upcoming challenges. It will be crucial to establish an education system that leverages traditional ways of thinking rather than working against them. This is entirely possible by using short-term incentive systems that lead to the long-term adjustment of our behavior and thinking to new demands. The previously cited neuroscientist from Frankfurt, Henning Beck, says that humanity changes its behavior long-term only if it receives an immediate reward.

This ties into what we discussed at the beginning. We are conditioned for short-term benefits because evolution gave decisive advantages over our competitors in the struggle for survival. In this way, we ensured the survival of our families and descendants. This pattern is deeply rooted in us and remains active today. Properly applied, however, it can also accelerate a shift in thinking.

We must also give "discipline" and "performance" greater importance again. It is right to protect our children (and ourselves) from excessive stress. If students in the Western world had to learn a complex writing system like Chinese characters, they would likely remain illiterate, because the Chinese script is complex and can only be mastered with discipline. However, if our children lose the basic ability to handle discipline and stress, our society loses capability.

There is no doubt that the latest findings from brain research should be considered that stress-free stimulation of our neural synapses increases learning ability. But there are things that must be done, even if they are not enjoyable. And this is only possible with discipline. If discipline is no longer part of learning, we lose a central ability. The same applies to performance: often, we are only pushed to achieve our best in competitive situations. Overcoming a competitor in a peaceful contest is one of the short-term rewards we are conditioned for. But the socially acceptable application must be learned.

Performance and competition are integral components of the market economy. Market economy and responsibility go hand in hand. Both must be committed to the common good, without descending into an overregulated common-good dictatorship. This is only possible with mutual tolerance; a tolerant market economy and a tolerant society. Tolerance, of course, does not mean failing to recognize performance. Without it, there is no progress.

Work-life balance is an important topic, no question. However, the Western world is in competition with Asia. Currently, neither Europe nor the U.S. is particularly well-positioned, and this competition will only increase. Already today, we are losing ground in this race. In school, our children are underchallenged. Later, at universities, we see the opposite situation. Here, curricula have become rigid, largely aimed at getting students into the workforce as quickly as possible, equipped only with the skills demanded by the economy. This is very stressful and leaves out important learning content crucial for life balance.

We must also, with an eye toward holistic innovation thinking, return to the idea of a new type of studium generale. Of course, knowledge has multiplied, making it impossible to cover even a fraction. However, it is important to develop an understanding of how disciplines in society, economy, and science are interconnected and influence each other, and to encourage a productive and inspiring engagement with science overall.

In all educational areas, the goal should be to foster an understanding of the positive impacts of innovation and to cultivate the desire to harness this potential optimally, not just individually, but for the benefit of society as a whole. This is the only way to maintain control over new technologies and how they are used. The real challenge will indeed be to create a mental state of openness to innovation early in the education cycle. But – and this repetition is intentional – we must not overlook older populations. Many older people are far more capable of learning than is generally assumed. The right incentives must be provided.

One example: for many years, I curated an annual innovation congress for a global trade association. The idea was to highlight new developments early, giving members the chance to become early adopters and pioneers. This group had expertise in "packaging" to offer compilation and visual presentation and the potential to lead in e-commerce. As early as 2015, we presented blockchain technology at our congress. Board support was zero. "What does this have to do with us?" a senior manager asked me. "We know our target audience. They all come to us after a certain age. Our business model will never change!" He was so wrong! The crisis hit the industry in the second year of COVID-19. Only four years after the conference, and just before the pandemic, a platform operator who had been one of the speakers contacted me. "Have you seen our latest update? We talked about it when I was on stage with you. Now we're doing it." The problem was that four years earlier, no one in the industry wanted to recognized the potential and thus no immediate advantage. They failed to seize the opportunity to position themselves early and remain market leaders. Now, young, non-industry startups or tech companies are taking the market from them.

We must create a fundamental mindset that anticipates the opportunities of innovation. You may recall the global mood after World War II in the 1950s and 1960s: there was broad hope that technology would provide all solutions. Many advances were made during this time. It was the era of the first moon landing, to cite just one example. Much nonsense was also done, from today's perspective. The dried-up Aral Sea resulted from Soviet-era cotton production, massively expanded through artificial irrigation. Nevertheless, we must move away from technological skepticism toward more optimism, without overdoing it. We will still make mistakes, but hopefully we can correct them faster and better.

Implementing this relatively simple approach into our education system is not easy. Our deeply rooted mindset is an obstacle. Everyone is already under enormous stress. I've already discussed society. Legislatures, in many places, try to manage the situation with a flood of specialized regulations, resulting in even more bureaucracy. In some countries, this has already led to a counter-movement: "de-bureaucratization with a chainsaw." Both approaches are only partially suitable for dealing with accelerated societal change. But perhaps this is just a cliché. Results in Argentina with their "chainsaw-approach" now make me cautiously optimistic.

Traditional education systems also suffer from outdated structures and the fact that many teaching staff are still trained and rooted in the analog world. This group also includes legislators and judges. Here lies a critical problem that requires a difficult, radical rethinking. Lifelong learning? In reality, there is little of it. Companies may offer employees specific courses tailored to

Resilience – Are we ready?

tasks. More often, however, older employees are still pushed out of companies because it is assumed, unspoken, that they are not flexible enough to adapt to new environments and younger employees are much cheaper. These decisions are often made by managers who themselves remain trapped in an analog mindset. Education must urgently address this. Lifelong learning is increasingly present at public universities, such as the Institute for Lifelong Learning (Institut de Formació Contínua-IL3) at the University of Barcelona, with its intergenerational offerings. Of course, these programs are still too narrow, too academic, and reach only a fraction of the target group. Promising approaches exist but need to be expanded and networked.

We need new, open systems that endure long-term and give people a sense of trust and security. I will only note a few points here. I advocate for open regulatory frameworks. General legal codes already define what is lawful and what is not. Why must I create new laws for every new area to fill a legally undefined space? And why must laws be repeatedly amended because, for example, new technical distribution methods are not mentioned, leaving the law inapplicable there? With accelerated change, lawmakers are already falling behind. Many laws are outdated by the time they come into effect and need revision. This is an endless cycle that only accelerates further, eventually overloading the legal system. Would it not suffice to say that the law applies to everything in this area, including unforeseen future applications? Legal practice would then be determined by jurisprudence. Rethinking is required here as well, to ensure created leeway is not narrowed again. This is precisely the problem. Traditional legal thinking is deeply rooted in the current generation. It hinders both innovation capability and acceptance. Change requires significant effort and progresses very slowly. This is also why establishing international legal standards is so difficult. The necessity is recognized by many, but rarely by lawmakers. They prefer familiar terrain or believe national interests must be protected.

Education must ultimately address this, not only in this discipline but everywhere. Despite the considerable time and material effort required, there is no alternative. Continuous education across all age groups is in the public interest. New approaches must urgently be generated and designed so that people are taken along on this journey into tomorrow. It has long been disproved that the ability to learn inevitably declines with age; however, genuine willingness to learn and curiosity must be awakened. For some, this never diminishes; for others, more effort is required to keep it alive. We have learned that this only succeeds if they can immediately expect an advantage for themselves and their families. Measures must also be "barrier-free," to adapt a current term. The easier access to offerings, the easier it is to engage with them.

The greatest challenge is adapting our education systems to new requirements. Traditionally, focus has been on educating children up to university. Many countries are making progress in early childhood education. Adult education, however, remains neglected in most systems, though positive developments are emerging. Existing professional development offerings, partly provided by companies and partly by public or private organizations, must be consolidated and significantly expanded. Dedicated courses are not even necessary. On YouTube, there are already very good tutorials on almost every topic. The best must be highlighted and the offerings made easily accessible. Broad visibility is needed. This new sector has enormous untapped potential. The goal must be to dispel fears and link them with positive expectations for change. How education is understood is closely tied to culture; therefore, universal recipes are difficult. Each cultural context must develop its own approach to create the necessary acceptance.

This obviously conflicts with the need for international standards, which are necessary in a globalized world to ensure equal opportunities. In higher education, some progress has been made, at least formally. Bachelor's and Master's programs are increasingly replacing regionally differing structures, creating comparability at the examination level. However, quality differences still hinder comparison. In adult education, apart from perhaps postgraduate programs from private or public universities focused solely on career advancement, there is still little. General promotion of future-readiness is largely absent, although a silver lining exists.

The question, of course, is who could sponsor such an initiative. Under the umbrella of the World Economic Forum (WEF) in Davos and the United Nations (UN) or their sub-organizations, interesting educational initiatives exist, largely in isolation, aiming in this direction. There are many small and large initiatives, including the DWI, that are active here and pursue similar approaches and goals. All of this should be expanded, networked, and amplified to gain more attention. I cannot claim to deliver ready-made solutions; the terrain is far too complex. Approaches and solutions must be developed collectively. I can, however, spark debate and raise awareness of the need.

Even outside the education sector, many tasks remain, not only individually but operationally. Administration and business must both contribute to helping citizens engage with innovations. This includes consistently reducing bureaucracy, which burdens both the economy and society – not merely transferring it to digital form, which worsens problems long-term. Administration and business also share responsibility. I speak of a renaissance of service expectations: easy access to digital services that make government offerings genuinely attractive online. Today, security barriers and the supposed protection of citizens' privacy often intimidate, especially non-digital natives.

Businesses are equally responsible, increasingly imposing services that were once standard, adding restrictive workflows that stress customers. Online shopping becomes complicated, and customers get lost in the app jungle. Customer satisfaction suffers, eventually lowering revenue. Costs for more service and easier digital access are quickly overcompensated. Let us finally fulfill the promise of digitization: to relieve people, not burden them further with unnecessary barriers. We accept only what we perceive as advantageous. Then willingness to embrace innovation rises. This may sound simple, but it is a crucial building block for greater innovation resilience.

As we have seen, tolerance is a prerequisite. Some may already wonder why this text is not gendered. This was a deliberate choice. Democracies, if taken seriously, are diverse, or in other words, tolerant. Anyone who wants to apply gendering may do so. The discussion it triggers is socially important and valid. But everyone must be free to decide. The problem is that advocates of diversity are often unwilling to grant the same tolerance they demand for themselves. This creates resentment and deepens divides that might otherwise not have been so pronounced.

Tolerance also lays the foundation for viable compromises, enabling us to tackle the ever-growing mountain of challenges ahead, whether at home or globally, whether technological, societal, climate-related, or otherwise.

Technological progress provides us with tools previously deemed impossible to transform our planet into the place we have always envisioned. Tolerance is one of the prerequisites for innovation resilience and thus the key to our future. It can only succeed through collective effort, based on understanding how we cognitively process things, and a positive attitude toward technological progress and divergent ways of life. This requires both collective and individual resilience, and openness to new possibilities.

If we seize this opportunity, we grasp it firmly. The road will be long, setbacks are inevitable, but in the end, we have the chance for a truly livable future. The alternative? A poisoned world full of fear, envy, and greed. When we consider this, the choice should be clear.

Resilience – Are we ready?

I do not claim to possess the sole salvific truth. What I present here are impulses, based on subjective analysis, even though I have tried to meet the aspiration of a broadened perspective. Some of it may be correct, much of it may overshoot, or be unpopular. I deliberately offer it for discussion. I know this debate will not be easy for me. If I demand painful compromises from all sides, I cannot exempt myself. As mentioned, at the Diplomatic World Institute we are developing the CAMPUS MUNDI platform, which will provide space for such discussions. The goal is clear: to contribute to offering real solutions. We are just one voice, but in chorus with other initiatives, we can be heard. Even if the perceived time for solutions grows ever shorter, we must take this path. Reason, responsibility, and tolerance are our guides.

By the same author:

Acknowledgments

Without support, a work like this could not be accomplished. My thanks go to Innovation Profiler Alexander Pinker, my partner on the video podcast Today & Tomorrow, who also consistently offers his advice and assistance beyond that. I am also grateful to our guests, some of whom are quoted here, for providing numerous insights, alternative perspectives, and sometimes simply for offering disagreement. I owe many impulses to Father Jörg Weinbach of the Deutsche Orden in Frankfurt, some of which have found their way into this book.

Barbara Dietrich deserves recurring mention, the tireless spirit behind the Diplomatic World project, which first gave me the space to develop my thoughts in this way. Markus Miksch deserves thanks for the insights into today's book publishing industry.

Of course, Marion deserves special acknowledgment, as at the end of each of my book projects. Without her support, not only of my writing, much in my life would simply not be possible.

By the same author:

Pandemia's Box

An Approach Towards a Sustainable Future for the Planet via Wholistic Innovation

Covid-19 changed our lives dramatically. This provides new risk but also new opportunities. We need to innovate not only in a technical sense but in every part of our lives, societies, the economy. It has an impact on our culture, finance, and health systems and even international relations. We need a wholistic approach to handle all this providing the chance to tackle the planet's burning problems and to come up with solutions faster and more sustainably. This book gives an idea of what "wholistic innovation" can be. The book tries not to provide a ready-made set of tools. The goal is to initiate a debate and a spark to inflame a process that in the end will provide these tools. The author takes a close look at the problems we are faced with in relevant sectors. The book provides many facts for each one and detailed web links opening the space for additional individual research.

DIPLOMATIC WORLD INSTITUTE Book Series Vol. ONE (2nd extended edition, ©2022)

Available on Amazon:

Paperback **ASIN :** B08XH2JL6M

e-book **ASIN :** B08WL9HC4J

About the author

Dieter Brockmeyer is an internationally recognized media and innovation expert. He is Chief Project Officer (CPO) of Diplomatic World, a media project that has been active since the turn of the millennium and, with its quarterly magazine, has established itself as an influential voice in the world of diplomacy in the Brussels area and far beyond.

He is cofounder and the innovation expert of the Diplomatic World Institute (DWI) in Brussels, as well as producer of the international video podcast series "*Today & Tomorrow* by

About the author

Diplomatic World", focusing on innovative thinking. He also curates international industry conferences and developed the Wholistic World Innovation Trophy, or *The TROPHY*, which the institute has celebrated annually each autumn in Barcelona since 2021.

The international award is based on the concept of "Wholistic Innovation," which he developed for the institute and explains in detail in his books *Pandemias Box* and *Campus*, published by DWI Publishing and available on Amazon. With *Resilience*, he now opens a new chapter, addressing an economic and societal challenge that remains widely underestimated.

© Dieter Brockmeyer

Brussels/Frankfurt 2026

All rights reserved

c/o: Diplomatic World Institute

Avenue Louise 146

1050 Brussels

Belgium

The Institute is a non-profit organization (VZW) under Belgian law, No.: 728 670 235

DWI Publishing is a unit within Diplomatic World Institute

ISBN: 979-8278-72-399-8

No part of this publication may be reproduced, stored in a retrieval system, or transmitted in any form or by any means, electronic, mechanical, photocopying, recording, or otherwise, without the prior written permission of the author/publisher.

First published in 2026

The publisher and author have made every effort to ensure the accuracy of the information contained in this book but assume no responsibility for errors or omissions.

Editor: Marion Witte

Cover Photo: Dieter Brockmeyer

Author's portrait: Nathalie Färber, Liquid Kommunikationsdesign, Frankfurt

Portrait of Barbara Dietrich: Private

Printed by Amazon Fulfillment (details last page)

www.ingramcontent.com/pod-product-compliance
Lightning Source LLC
Chambersburg PA
CBHW050103230526
45470CB00004B/1660